WORLD HEALTH ORGANIZATION

INTERNATIONAL AGENCY FOR RESEARCH ON CANCER

STATISTICAL METHODS IN CANCER RESEARCH

VOLUME III – The design and analysis of long-term animal experiments

by

J.J. GART, D. KREWSKI, P.N. LEE, R.E. TARONE & J. WAHRENDORF

IARC Scientific Publications No. 79

INTERNATIONAL AGENCY FOR RESEARCH ON CANCER
LYON
1986

The International Agency for Research on Cancer (IARC) was established in 1965 by the World Health Assembly, as an independently financed organization within the framework of the World Health Organization. The headquarters of the Agency are at Lyon, France.

The Agency conducts a programme of research concentrating particularly on the epidemiology of cancer and the study of potential carcinogens in the human environment. Its field studies are supplemented by biological and chemical research carried out in the Agency's laboratories in Lyon and, through collaborative research agreements, in national research institutions in many countries. The Agency also conducts a programme for the education and training of personnel for cancer research.

The publications of the Agency are intended to contribute to the dissemination of authoritative information on different aspects of cancer research.

Distributed for the International Agency for Research on Cancer
by Oxford University Press, Walton Street, Oxford OX2 6DP, UK

London   New York   Toronto
Delhi   Bombay   Calcutta   Madras   Karachi
Kuala Lumpur   Singapore   Hong Kong   Tokyo
Nairobi   Dar es Salaam   Cape Town
Melbourne   Auckland

Oxford is a trade mark of Oxford University Press

Distributed in the United States
by Oxford University Press, New York

ISBN 92 832 1179 0
ISSN 0300–5085

© International Agency for Research on Cancer 1986
150 cours Albert Thomas, 69372 Lyon Cedex 08, France

Filmset in Northern Ireland by The Universities Press (Belfast) Ltd.

CONTENTS

FOREWORD

In 1980, the Agency started a series on 'Statistical Methods in Cancer Research' within its programme of scientific publications. The first volume was devoted to the analysis of case-control studies and received wide attention as the first comprehensive textbook on this subject. It demonstrated clearly the need for authoritative texts in the field of cancer biostatistics, in which there have been considerable methodological developments in the last few decades.

The present volume, on the design and analysis of long-term animal experiments, addresses an area of cancer research that is of great importance to the understanding of carcinogenic risk factors. Long-term animal experiments play a major role in assessment of the carcinogenicity of chemicals in the light of the methodological difficulties of some epidemiological studies and the biological limitations of short-term assays. An effort to standardize methods for long-term tests has been made in recent years, and a thorough treatment of the statistical issues involved was therefore very necessary. The publication of this volume is in concordance with the Agency's aim to promote all facets of cancer research by disseminating information and providing methodological support.

It is a pleasure to commend the five authors on this book, which should prove to be of value for biostatisticians working in the field of long-term animal experiments as well as for experimentalists who want to learn about the statistical concepts involved and all those who have to interpret findings from such studies for regulatory purposes. Last but not least, it may also serve as a textbook for students in biostatistics and related fields.

The Agency is, at the same time, actively pursuing research into developing tests that can be used to make valid predictions of carcinogenic risk factors but which do not necessitate the use of experimental animals.

L. Tomatis, MD
Director, IARC

PREFACE

The roots of this monograph on the design and analysis of long-term animal experiments can be found in the programme of the *IARC Monographs on the Evaluation of the Carcinogenic Risk of Chemicals to Humans,* which started in 1971. Since adequate data for humans were often not available, this programme had to base its evaluations predominantly on data from long-term carcinogenicity studies in animals. In the event that there was sufficient evidence for a chemical to be considered carcinogenic in animals, it was noted that this compound should be regarded, for practical purposes, as if it represented a carcinogenic risk to humans. Also, when human data were available, animal studies provided, and continue to provide, valuable additional information for a chemical's evaluation. The Agency recognized that long-term, and, indeed, also short-term assays, could play their role in screening for carcinogens most effectively only if the methodology was thoroughly understood and standardized. Therefore, a meeting was organized in 1979 which resulted in the publication of a critical appraisal of the methodological aspects of long-term and short-term assays (IARC, 1980). During this meeting it became apparent that statistical issues play an important role in assessing the results of long-term carcinogenicity studies.

In order to convey the essential aspects of the statistical analysis of long-term carcinogenicity studies, Richard Peto initiated the writing of an Annex giving guidelines for simple, sensitive significance tests for carcinogenic effects in long-term animal experiments (Peto *et al.,* 1980). This Annex proved to be a very valuable text not only for experimentalists but also for statisticians who had to apply these methods or were developing the methodology further.

Statistical methods for long-term animal experiments have to account for the unique biological and toxicological problems encountered in animal tests and the development of such methods has become an active area of research. To summarize the recent research in a monograph was therefore felt to be a very natural consequence of the publication of the Annex and to fit very well in the Agency's series of monographs on *Statistical Methods in Cancer Research.*

At a first meeting held in Lyon in 1981, we outlined the structure of this book and organized its production; drafts were subsequently prepared and circulated, comments from colleagues were obtained, and further meetings and revisions followed before the final product became available.

The book is intended to be of interest not only for biostatisticians but also for other scientists involved in the design, conduct and interpretation of long-term animal experiments. The first three chapters are written in a completely nontechnical fashion in order to introduce clearly the fundamental aspects of long-term animal experiments and to identify the issues involved in the design of such studies. The statistical methods presented in the later chapters are illustrated using data from real experiments. For teaching purposes, we also give a shorter, constructed example for which the relevant calculations can be done easily with a pocket calculator in the classroom. A detailed outline of the monograph is given at the end of the first chapter.

The book has profited considerably from the stimulating background provided by the *IARC Monographs on the Evaluation of the Carcinogenic Risk of Chemicals to Humans*. Many colleagues have given valuable comments to improve earlier drafts and assisted in a careful final reading; we gratefully acknowledge the contributions of: N.E. Breslow, C.C. Brown, J.R.P. Cabral, D. Colin, N.E. Day, A. Dewanji, G.E. Dinse, L. Edler, R.A. Griesemer, J.K. Haseman, W. Lehmacher, B. McKnight, D. Murdoch, C. Portier, M. Schumacher, J.D. Wilbourn and T. Yanagimoto.

We are extremely grateful to Annick Rivoire for her skilful and untiring efforts in guiding the manuscript through its various phases of development at IARC. We would also like to thank Doris Foster and Judy Graham for their bibliographic assistance at Health and Welfare Canada.

Finally, we decided to arrange the authors' names in alphabetical order.

J.J. Gart,[1] D. Krewski,[2] P.N. Lee,[3] R.E. Tarone[1] and J. Wahrendorf[4]

---

[1] Mathematical Statistics and Applied Mathematics Section, Biostatistics Branch, National Cancer Institute, Bethesda, MD 20892, USA.

[2] Biostatistics and Computer Applications, Health and Welfare Canada, Tunney's Pasture, Ottawa, Ontario K1A OL2, Canada and Department of Mathematics and Statistics, Carleton University, Ottawa, Ontario, Canada.

[3] P.N. Lee Statistics and Computing Ltd, 25 Cedar Road, Sutton, Surrey SM2 5DG, UK.

[4] Unit of Biostatistics and Field Studies, International Agency for Research on Cancer, 150 cours Albert-Thomas, 69372 Lyon Cedex 08, France; present address: German Cancer Research Center, Im Neuenheimer Feld 280, 6900 Heidelberg, Federal Republic of Germany.

1. GENERAL INTRODUCTION AND OUTLINE

CHAPTER 1

GENERAL INTRODUCTION AND OUTLINE

Animal experimentation in cancer research dates back to 1775, when Peyrithe injected human cancerous material into a dog in an attempt to induce neoplasms (Shimkin, 1977). Since then, many attempts have been made to transplant human tumours to animals, and the first successful experiment of this type was reported by Nowinsky (1876). The first use of animals in the field of chemical carcinogenesis is attributed to Yamagiwa and Ichikawa (1915). By the application of crude coal-tar dissolved in benzene to the ears of 137 rabbits every two or three days, they induced 'folliculo epitheliomas' in nearly all ears after a period longer than 100 days, and, within a year, seven cases of fully developed invasive carcinomas occurred. The relevance of animal carcinogenesis experiments to man was demonstrated by the induction of bladder tumours in dogs exposed to aromatic amines (Hueper & Wolfe, 1937; Hueper *et al.*, 1938); these results corresponded to epidemiological observations that workers exposed to similar agents in the workplace had an increased occurrence of bladder neoplasms.

Since then, long-term animal experiments have become an integral part of cancer research. They are used for various purposes, including screening chemicals or other exposures (for example, ionizing radiation) for their carcinogenic potential and for elucidating the mechanisms of carcinogenic action of known cancer-causing agents. Their importance in the process of assessing cancer risks has been discussed repeatedly (see, for example, Tomatis, 1977, 1979; Weinstein, 1981), and even with the increasing use of short-term mutagenicity assays, animal experiments remain an indispensable component of cancer risk assessment.

A review of the *IARC Monographs on the Evaluation of the Carcinogenic Risk of Chemicals to Humans* indicates that in the first 32 volumes published between 1971 and 1983, 654 chemicals, groups of chemicals, industrial processes or occupational exposures were considered. Epidemiological data were available for only 157 of these; an assessment of these data led to the classification as carcinogenic to humans of seven industrial processes or occupational exposures and 23 chemicals or groups of chemicals. However, considerably more data on the cancer risk of these 654 exposures were available from long-term animal experiments. Based on animal data, there was sufficient evidence of carcinogenicity for 187 chemicals, limited evidence of carcinogenicity for 176 chemicals, no evidence of carcinogenicity for eight chemicals and the data were inadequate to evaluate 260 chemicals. Most known human carcinogens have been found to be carcinogenic in experimental animals by appropriate testing, and in many cases the experimental evidence preceded the epidemiologic evidence of human risk.

The use of long-term animal experiments for routine testing of chemicals has become part of the procedure for setting public health regulations in many countries. The conduct of carcinogenicity studies is often required for the registration of new compounds, and compounds already in use are being progressively considered for screening. For example, in the USA the Bioassay Program of the National Cancer Institute investigated several hundreds of compounds, and further studies are now in progress under the National Toxicology Program with about 25 long-term car-cinogenicity studies started every year. It is difficult to obtain an exact number of such studies undertaken worldwide each year, but an estimate of the order of magnitude of several hundred may be appropriate. In the *Information Bulletin on the Survey of Chemicals Being Tested for Carcinogenicity* (IARC, 1982a), about 1000 chemicals were reported to be under test in 103 institutes in 16 countries.

A number of methodological improvements have led to the establishment of long-term animal experiments as an indispensable part of cancer research. Four areas of improvement, in particular, may be identified. Many authors view the control of genetic variability by the introduction of inbred strains as a progressive step in biomedical research (for an interesting review, see Festing, 1979). The control of genetic variability allows for the reproducibility of the experiments, an important criterion in experimental science.

Advances have also been made in the design or conduct of the experiments. Minimal requirements for an acceptable protocol have been established for several aspects of carcinogen bioassay experiments; these include the number and size of the experimental groups, the selection of dose and route of administration, randomization, animal husbandry, diet, duration of exposure and observation period (Interagency Regulatory Liaison Group, 1979; IARC, 1980). Currently, the role of diet in long-term animal experiments is a key topic for discussion.

Thirdly, the pathological evaluation of experimental animals has undergone con-siderable change. Careful macroscopic inspections, standardized histology and broad pathological reviews have contributed to improvements in the quality of pathological data (Interagency Regulatory Liaison Group, 1979). The experimental pathologist's work may be organized very differently in different settings, depending on the size of the laboratory, the design of the experiments and related aspects, which will be discussed in Chapter 3. Whereas some pathologists observe the animals from the beginning of the experiment to the final evaluation, thus being familiar with all clinical signs, others conduct the final review without such information. The question as to whether pathological reviews should be conducted blindly, that is, without knowledge of an animal's treatment group, has been a very controversial subject which remains not fully resolved (Weinberger, 1973; Fears & Schneiderman, 1974; Ward et al., 1978).

Finally, statistical methods used to analyse such experiments have improved considerably. With the realization that differences in intercurrent mortality have to be adjusted for in the comparison of proportions of tumour-bearing animals, and the development of specific models to describe various aspects of the carcinogenic process, the range of statistical methods available has expanded greatly.

The purpose of this book is to provide a comprehensive introduction into all statistical aspects of long-term animal experiments. A variety of methods published recently in the scientific literature are summarized comprehensively. This is intended mainly to help statisticians working in the field of long-term animal experiments, as well as helping experimentalists with an interest in the quantitative interpretation of their results to familiarize themselves with the methods available. However, as the first three chapters provide a nontechnical introduction to major concepts related to the statistical evaluation of long-term animal studies, the book will be of interest to a much wider audience of scientists, public health officers and others who are concerned with long-term animal studies. Finally, it can also serve as a text for courses taught to students in biostatistics or a related medical field.

Throughout the book, emphasis is placed on the occurrence of tumours as the basic endpoint in long-term animal experiments. However, some attention is also given to the many other observations that have to be taken into consideration for the statistical assessment of a carcinogenic effect.

With these aims, this book complements Supplement 2 to the *IARC Monographs* series, entitled 'Long-term and short term screening assays for carcinogens: A critical appraisal'. That publication includes a section entitled 'Basic requirements for long-term assays for carcinogenicity' and an Annex, 'Guidelines for simple, sensitive significance tests for carcinogenic effects in long-term animal experiments'. Whereas the section on basic requirements mainly addresses the conduct of the experiments, the annex on statistical guidelines describes in non-technical terms one method for routine analyses of carcinogenicity experiments. This method is also described in this monograph, where it is included in a general and more technical description of the methods available. In addition, the current monograph discusses the circumstances under which the various methods are suitable and the assumptions that underlie them.

The contents of the remaining chapters are as follows. In the second chapter, general considerations related to the statistical evaluation of long-term animal experiments are discussed. The third chapter deals with various aspects of the experimental design. In the fourth chapter, some illustrative data sets are introduced which serve throughout the monograph as examples. The fifth chapter then discusses methods available for comparing tumour occurrence in different experimental groups. The sixth chapter outlines the fitting of statistical models, which are useful in summarizing patterns of tumour incidence and in attempting to elucidate mechanisms of action. The seventh chapter is devoted to a list of special topics, all of which are important in this field and deserve careful attention; they include multiple statistical comparisons, multiplicity of tumours, litter effects, association among tumour types, the use of historical controls and multigenerational studies. Finally, in the eighth chapter, we deal with the analysis of concomitant observations such as survival rates, body weight and food consumption, clinical signs of toxicity, haematological variables, and organ weight.

2. GENERAL CONSIDERATIONS ON THE EVALUATION OF ANIMAL CARCINOGENESIS STUDIES

2.1 Introduction

2.2 Determination of relevant biological events

2.3 Non-neoplastic precursors of a neoplasm

2.4 Adjustment for intercurrent mortality

2.5 Combining analyses as opposed to pooling data

2.6 Considerations related to sex and species

2.7 Consideration of tumour dose-response

2.8 Observable tumours

2.9 Problems with occult tumour data

2.10 Contexts of observation for occult tumours

2.11 Potential for bias in the analysis of fatal tumours

2.12 Potential for bias in the analysis of incidental tumours

2.13 Combining analyses of fatal and incidental tumour data

2.14 Context of observation unavailable

2.15 Analysis of crude tumour rates

2.16 Concomitant information

2.17 Need for interdisciplinary decision process

2.18 Overall evaluation of carcinogenicity

CHAPTER 2

GENERAL CONSIDERATIONS ON THE EVALUATION OF ANIMAL CARCINOGENESIS EXPERIMENTS

2.1 Introduction

The primary purpose of a long-term carcinogenicity experiment is to determine if the administration of a test substance to animals of some species alters the normal pattern of tumour development in that species. In a typical long-term carcinogenicity experiment, a pool of animals is divided by randomization into several groups. One group serves as a concurrent control group, while the remaining groups are exposed to various dose levels of the test substance by some appropriate route of administration. The test animals are observed for a major portion of their lifespan, and all animals that die during the study are subjected to necropsy unless they are substantially cannibalized or autolysed. The experiment is terminated according to a predetermined stopping rule, for example, after a fixed period of time on study, or when mortality in the control or lowest-dose group exceeds a specified limit such as 50% (see IARC, 1980). At termination, all surviving animals are killed and subjected to necropsy. For each animal, tissues are taken from several organ sites and examined histopathologically. The basic data obtained from each animal are the times of appearance of any visible tumours, the time of death, the cause of death (in so far as the cause can be determined), a list of organs examined at necropsy, and the histopathological diagnoses for those organs examined. The goal of the statistical analysis of these survival and pathology data is to quantify the strength of evidence regarding the carcinogenic potential of the test agent. In this book the emphasis will be on evaluating the carcinogenicity of a test substance only in the species under study. For recent reports on the role of long-term animal tests in assessing the potential carcinogenic risk to man, see Weisburger and Williams (1981), Squire (1981), IARC (1982b), and Interdisciplinary Panel on Carcinogenicity (1984), as well as the remarks in Section 2.18.

2.2 Determination of relevant biological events

In long-term carcinogenicity studies, the experimental outcome of interest is the occurrence of a tumour at some target organ. Throughout this book, the word 'tumour' will be used quite generally to refer to a well-defined class of neoplastic lesions. The determination of which class of lesions at a given organ should be analysed for evidence of carcinogenicity can be extremely difficult and requires the judgement of experienced pathologists. Such determinations will vary from species to species and among organs within a species, but generally it is advisable to restrict the grouping of lesions to tumours of the same histological type arising in the same type of tissue (IARC, 1980). In addition, analyses should be restricted to primary tumours rather than to a grouping of primary and metastatic tumours. The importance of and difficulty in defining the class of lesions which represents a carcinogenic response indicates the need for a close working relationship between pathologists and statisticians involved in the evaluation of animal carcinogenesis studies.

Although evidence of a carcinogenic effect can occur in any of the several organs examined, the effect of a carcinogenic agent is likely to be concentrated in one or a few target organs. A clear-cut carcinogenic response at one target organ may be obscured by an analysis based on the incidence of all tumours, regardless of their sites of occurrence. This is particularly true in species which have a high rate of naturally occurring tumours. Thus, unless there is information available prior to the evaluation of an experiment indicating that a test agent is likely to have carcinogenic potential at more than one organ or tissue type, the pooling of tumour incidence data from two or more sites should be avoided. In general, statistical analyses of tumour incidence data should be restricted to tumours which develop at a specific organ, or to tumours of a type known to have a multicentric origin (for example, leukaemias) (IARC, 1980). In exceptional cases, it may be reasonable to pool tumours at biologically related sites or tumours with common morphological characteristics, but this should be done only in close consultation with a pathologist.

The evaluation of a long-term carcinogenicity experiment is complicated by the fact that there is no single well-defined biological response that characterizes a carcinogenic effect. A carcinogenic agent may cause any of several types of alterations in the normal pattern of tumour development, and data from a carcinogenicity experiment must be examined carefully for each of the possible changes. The most important carcinogenic response is an increase in age-specific rates of tumour incidence in exposed animals over some portion of the lifespan of the test species, leading to an increased lifetime probability of developing a tumour. Another possible carcinogenic response is an acceleration of tumour development in exposed groups; that is, tumours similar to those occurring naturally may develop earlier in life in exposed animals than in control animals, while the lifetime probability of developing a tumour remains unchanged. The possibility of such an acceleration effect, which might arise if only a fixed proportion of the test animals are at risk of developing a tumour, has been proposed (Lagakos & Mosteller, 1981); however, there is little experimental evidence for such an effect in inbred strains. For some species, an increase in the number of tumours at some organ in exposed animals may indicate carcinogenicity (Tomatis et al., 1973; Shimkin & Stoner, 1975; Ward & Weisburger, 1975; Peraino et al., 1977; Ward & Vlahakis, 1978). Potential problems in the interpretation of data on tumour multiplicity are discussed by Peto et al. (1980). Qualitative morphological or biological differences between tumours found in exposed animals and tumours found in control animals may also provide evidence of a carcinogenic effect. For example, tumours that have unusual morphology may be observed in exposed animals but not in control animals, or tumours in exposed animals may be more prone to metastasize than tumours in control animals (Frith et al., 1979; Reznik & Ward, 1979; Hoover et al., 1980; Stinson et al., 1981). In many cases, the identification of unusual patterns of tumour development depends heavily on the knowledge and judgement of pathologists and toxicologists (Ward, 1984).

Although the statistical analysis of tumour and survival data is an essential component of the evaluation of carcinogenicity, care must be taken, in assessing carcinogenic potential, not to place total reliance on the finding of statistically significant results. For example, there may be minor changes in the normal tumour pattern, which might not be detected by standard statistical methods, but may nonetheless be evidence of carcinogenicity. Of particular importance is the possibility that a few exposed animals, but no control animal, may develop a tumour at an organ in which tumours very rarely occur naturally. Historical control data can sometimes be used to increase the sensitivity of statistical tests for rare tumours (see Chapter 7). The induction of tumours that are normally rare may be extremely important in assessing human risk (Squire, 1981). On the other hand, even a significant tumour increase in only one sex at an organ in which naturally occurring tumours are quite common may not provide convincing evidence of carcinogenicity in the test species (Fears *et al.*, 1977; Gart *et al.*, 1979). The evaluation of carcinogenicity in these and other cases may rest as much on biological, toxicological or pharmacological considerations as on the presence or absence of statistical significance.

2.3 Non-neoplastic precursors of a neoplasm

Although the emphasis in ensuing chapters is on the analysis of tumour data, it should be noted that evidence of carcinogenicity can also be obtained from data on non-neoplastic lesions, when such lesions are precursors of a neoplasm. When both neoplastic lesions and non-neoplastic precursors are discovered in animals in the same experiment, analyses of the non-neoplastic precursors separate from the neoplastic lesions must be interpreted with care. If a test agent is extremely efficient at converting non-neoplastic precursors to neoplasms, an analysis based solely on non-neoplastic lesions may show a dose-related decrease, while an analysis of neoplastic lesions shows a dose-related increase, indicating that the compound is carcinogenic. Two analyses may be informative in such a situation, one based only on animals with neoplastic lesions and the other based on animals with either a neoplasm or a non-neoplastic precursor.

2.4 Adjustment for intercurrent mortality

Identification of differences in overall mortality among exposure groups is an important step in evaluating a carcinogen bioassay. Even if a single biological response characterizes a carcinogenic effect, the analysis of changes in the normal pattern of tumour development may still be difficult. This is because the observed experimental outcome corresponding to a particular alteration in tumour development can vary from experiment to experiment, depending on whether the control and exposed groups differ in intercurrent mortality. Intercurrent mortality refers to interim deaths not related to the development of the particular type or class of tumours to be analysed for evidence of carcinogenicity (Peto *et al.*, 1980). Adjusting for such intercurrent mortality is an important consideration in evaluating a carcinogenesis experiment. Consider an agent which causes increases in the underlying (and in general, unobservable) age-specific rates of tumour incidence throughout the lifespan of the test species. If intercurrent mortality rates are equal in the control and exposed groups, then the increased age-specific rate of tumour incidence in the exposed animals results in an increase in the observed proportion of exposed animals that develop tumours during the experiment. If, however, the intercurrent mortality rates are higher in exposed animals than control animals, then the observed proportion of exposed animals that develop tumours may not be increased, in spite of underlying age-specific rates of tumour incidence that are uniformly higher.

To illustrate this point, suppose that the test agent is quite toxic and that the agent induces tumours that do not shorten the lives of tumour-bearing animals. In a hypothetical experiment summarized in Table 2.1 with 100 animals exposed to the test

Table 2.1 Proportions of control and exposed animals dying with a tumour in early and late stages of a hypothetical carcinogenesis experiment (the denominator is the number of animals dying in each time period, and the numerator is the number of these dead animals in which tumours were found at necropsy)

	Control	Exposed
Died prior to 15 months	1/20 (5%)	18/90 (20%)
Died after 15 months	24/80 (30%)	7/10 (70%)
Total for experiment	25/100 (25%)	25/100 (25%)

agent and 100 unexposed control animals, suppose that, in the first 15 months, 90 exposed animals die and 18 of these exposed animals are discovered to have tumours, while only 20 control animals die and one of these control animals is discovered to have a tumour. Thus, during the first 15 months of the experiment, the observed percentage of animals with tumour is 20% in the exposed group and 5% in the control group. Suppose now that seven of the remaining ten exposed animals and 24 of the remaining 80 control animals develop tumours in the last months of the experiment, so that the observed percentage of animals with tumour among those surviving 16 months or longer is 70% in the exposed group and 30% in the control group. In spite of the fact that the proportion of exposed animals with tumour is higher in both the early and late stages of the experiment, the percentage of animals that are observed with tumour in the entire experiment is 25% (25/100) in both the exposed and control group. Since it is common in animal carcinogenesis experiments for exposed animals to have higher intercurrent mortality rates than control animals, and for prevalence rates in both the control and exposed groups to increase as an experiment progresses, this hypothetical example illustrates an outcome that can occur in practice. It is clear that, in the analyses of any changes in the pattern of tumour development, differences in longevity between control and exposed animals must be taken into consideration.

2.5   Combining analyses as opposed to pooling data

The hypothetical example given in Table 2.1 illustrates an important source of bias that can arise in the analysis of animal carcinogenicity experiments. This is the inappropriate pooling of data. Data pooling refers to the practice of calculating summary tumour rates by simply adding the number of animals with tumour in several distinct strata (for example, two time periods in the above example) and dividing this sum by the total number of animals (in all strata). The evaluation of the carcinogenicity of a test agent often involves the analysis of data from several strata. Some strata are naturally defined; for example, each sex of each species of test animal defines a stratum. Within each sex/species experiment, further stratification is possible, such as the subdivision of data by time period in Table 2.1. The spontaneous tumour rate will usually vary from stratum to stratum (this being the motivation for the stratification by time period in Table 2.1); however, under the null hypothesis that the test agent has no carcinogenic effect, the tumour rates within each stratum for all exposure groups should be equal. The example of Table 2.1 illustrates the danger of pooling data from different strata prior to statistical analysis. An effect which is present in all strata may be obscured when data from several strata are pooled. Analysis of data from each time period in Table 2.1 would show up the increased tumour rates in exposed animals in both strata. By using appropriate methods, analyses from several strata can be combined. Increases or decreases in tumour incidence in all strata provide enhanced statistical evidence regarding the carcinogenic potential of the test agent. This combining of analyses is preferable to analysing pooled data, and appropriate methods for combining analyses are discussed in Chapter 5.

2.6   Considerations related to sex and species

Typically, animal carcinogenesis experiments are carried out using both sexes of each test species. In the statistical analysis of long-term carcinogenicity tests, tumour incidence data from male and female animals should never be pooled. Because there are often large differences between sexes in the rates of naturally occurring tumours, such pooling of data from both sexes can lead to incorrect inferences (e.g., see Gart, 1962). In addition, hormonal differences may lead to carcinogenic risk in only one sex or to a substantially higher risk in one sex. Thus, the data for each sex should be analysed separately. The demonstration of a similar carcinogenic effect in both sexes of a species strengthens the scientific inference regarding carcinogenicity and can help rule out the possibility of a spurious (for example, false-positive) result. Where appropriate (that is, when a similar carcinogenic effect is observed in both sexes), the separate statistical analyses can be combined to obtain a summary quantification of risk in the test species.

An analogous situation arises if a compound has been tested in multiple experiments using different species or strains of test animals. Because of genetic differences, there can be considerable variability among species and among strains within a species with respect to their rates of naturally occurring tumours and their susceptibility to compound-induced tumours (Haseman & Hoel, 1979). Thus, as is the case with data from both sexes within a species, data from different species and strains should never be pooled. The data from each species or strain should be analysed and evaluated separately. The finding of a carcinogenic effect in more than one species or strain strengthens the scientific inference, particularly if the carcinogenic effect is at the same organ or of the same histological type in different species. If similar patterns of tumour induction are observed in multiple species, then the separate statistical analyses can be combined across species to obtain a summary quantification of risk for the test compound.

2.7 Consideration of tumour dose-response

In many animal carcinogenesis experiments, groups of animals are exposed at multiple dose levels of a test substance. If the test substance is carcinogenic, one would, in most situations, expect tumour rates to increase with increasing dose level. If a test substance protects against tumour induction or development, one would, in most situations, expect tumour rates to decrease with increasing dose level. Certainly, a monotonic change in observed tumour rates with increasing dose should strengthen the inference that differences in tumour rates are due to exposure to the test substance, with steeper dose-response curves providing stronger evidence of an effect. Thus, in choosing statistical methods, priority should be given to methods that are more powerful (that is, are more likely to indicate statistical significance) when observed tumour rates increase monotonically with dose. Accordingly, in presenting methods for analysing multiple-dose experiments, emphasis will be on testing for monotonic trends in tumour rates rather than on more general heterogeneity tests. In interpreting the results of such trend tests, it should be noted that a significant trend does not necessarily imply increased cancer risk at very low doses. Also, in the presence of a significant trend test, the lack of an increase in observed tumour rates at lowest dose levels should not be taken as evidence of a threshold. With the small numbers of animals typically used in animal carcinogenesis experiments, reasonable inference can seldom be made regarding the shape of the tumour-response curve at very low doses (Portier & Hoel, 1983a).

2.8 Observable tumours

For certain tissues such as the skin, tumours are visible, and hence their development can be closely monitored. There are other organs in which the presence of internal tumours may be evident before the death of the tumour-bearing animals; for example, mammary tumours often can be identified by palpation in living rodents (Davis *et al.*, 1956). Once an observational endpoint has been defined clearly in such cases (for example, the occurrence of a skin tumour of some prespecified minimum diameter), then the age at which an animal obtains this endpoint can be observed relatively unambiguously. Such tumours have been termed mortality-independent tumours (Peto *et al.*, 1980), because their observation does not require the death of the tumour-bearing animals. Whether or not a given animal develops such a tumour, however, is dependent upon mortality. An animal that dies at an early age is less likely to have developed such a tumour than is an animal that survives to old age. Thus, in subsequent discussions, these tumours will be referred to as 'observable' rather than 'mortality-independent'. More specifically, tumours of a certain type will be referred to as observable if, when they reach some standard point in their development, they can be identified in all living animals. The class of observable tumours includes those such as palpable tumours, which are not necessarily visible. For observable tumours, incidence data can be analysed using standard life-table methods (Peto *et al.,* 1980). These life-table methods test for earlier or more frequent observation of tumours in exposed groups while correcting for differences in intercurrent mortality rates. The interpretation of the life-table analyses for observable tumours is relatively straightforward; significant differences among control and exposed groups provide evidence of agent-related changes in tumour onset or development time or in the magnitude of age-specific rates of tumour incidence.

2.9   Problems with occult tumour data

Internal tumours that can be discovered only at necropsy are termed 'occult' tumours. Perhaps the most difficult aspect of the statistical analysis of data from a long-term carcinogenicity test is that inferences must be made concerning unobservable quantities, the ages at onset of occult tumours, when all that actually can be observed are the ages at death of the tumour-bearing animals. Whereas the analysis of observable tumours is relatively straightforward, analysis of occult tumours involves additional assumptions concerning the relationship between the observable outcome, age at death with a tumour, and the endpoint for which it is a surrogate, namely, age at onset of the tumour. Even with the additional assumptions, it is not, in general, possible to test directly hypotheses about the rates of tumour onset, that is, tumour incidence rates (McKnight & Crowley, 1984). It will be important to keep in mind throughout the following discussion the surrogate role played by age at death with tumour.

2.10   Contexts of observation for occult tumours

For occult tumours, the statistical analysis is complicated by the dependence of the observation of these tumours on the death of the animals. Differences in the ages at which particular tumours are observed can result from changes in age-specific tumour incidence or mortality rates or from changes in tumour growth rates, but they can also result from changes in age-specific mortality rates associated with causes unrelated to the development of that particular type of tumour. Thus, the early appearance of tumours in dying exposed animals may or may not provide evidence of carcinogenicity. If tumours at some organ are killing their host animals, the observation of tumours earlier in exposed groups than in the control group may be evidence of carcinogenicity of the test agent. If, on the other hand, tumours at a particular target organ are occurring with equal age-specific incidence rates in control and exposed groups, while animals are dying younger in exposed groups due to causes unrelated to these tumours (for example, toxicity or tumours at another site), then the observation of these tumours earlier in exposed groups should not be considered as evidence of carcinogenicity at the target organ. The early appearance of tumours in such cases may reflect simply the higher intercurrent mortality rate in the exposed groups. Thus, it has been noted that the context of the observation of a tumour should be determined as accurately as possible (Hoel & Walburg, 1972; Peto *et al.*, 1980). Accordingly, in subsequent discussions, a tumour which either directly or indirectly kills its host will be said to have been observed in a fatal context. A tumour which is observed at necropsy of an animal which has died of some unrelated cause will be said to have been observed in an incidental context. There are many difficulties involved in determining the context of observation of a tumour (Gart, 1975; Gart *et al.*, 1979; Peto *et al.*, 1980); however, when such determinations cannot be made, the range of appropriate statistical methods and the interpretation of the statistical analysis may be limited. Thus, when possible, the effort should be made to determine the context of observation of each tumour (Peto *et al.*, 1980). It is important that such determinations be accurate, as errors can lead to bias in the statistical analysis (Lagakos, 1982).

2.11 Potential for bias in the analysis of fatal tumours

Even if the contexts of observation have been accurately determined, there still can be difficulties in the statistical analysis of occult tumours. The analysis of data on fatal tumours is based on the observed ages at which animals are killed by the tumours, and makes use of the same life-table methods used in the analysis of data on tumours that are observable. Significant differences between control and exposed groups are taken as evidence that the test agent is associated with changes in the age-specific rates of death caused by tumour (Peto *et al.*, 1980). As noted earlier, the observation, age at death caused by tumour, is used as a surrogate for a variable that cannot be observed, namely, either the age at onset of tumour or the age at which a tumour reaches some standard developmental stage. If a tumour is rapidly lethal, then, in fact, the age at death may closely approximate the age at onset. In classifying a tumour as occurring in a fatal context, however, no distinction is made between tumours which kill their hosts rapidly after onset and tumours which kill their hosts slowly, perhaps several months after onset. In interpreting the analysis of tumours observed in a fatal context, it should be considered that earlier deaths due to tumours do not necessarily imply earlier onset times or more rapid development of tumours, particularly in cases in which the tumours are killing their hosts slowly.

In the case of observable tumours which can be detected before they become life-threatening, the interpretation of the life-table analysis in terms of tumour development is relatively straightforward. However, the interpretation of fatal tumour analyses in terms of tumour development requires the assumption that control and exposed animals are equally likely to be killed by a tumour at any particular stage in the tumour's development. This assumption may not always be reasonable. Consider an experiment in which animals in an exposed group have age-specific incidence rates for some tumours that are identical to those for the control animals. Suppose that these tumours are all observed in a fatal context, but that the exposed animals die more rapidly than the control animals once they develop a tumour. This acceleration of deaths due to tumour is not necessarily evidence of enhanced tumour development in the exposed animals. In many chemical carcinogenesis experiments, exposed animals are administered a test compound at the maximum tolerated dose – a high dose which may place the exposed animals under great physiological stress. It is possible that a tumour might kill such a stressed animal at an early stage in the tumour's development, a stage at which the tumour would be unlikely to kill a healthier control animal. Of course, if age-specific incidence rates are unaffected by the test agent, then the proportion of exposed animals observed with tumours during the entire experiment should not be higher than the proportion of control animals observed with tumours. Accordingly, analyses of fatal tumours which indicate a significant acceleration of death caused by tumour, without an accompanying increase in the proportion of animals observed with tumours, must be interpreted with care (Lagakos & Mosteller, 1981; Mantel, 1980).

2.12   Potential for bias in the analysis of incidental tumours

As in the case of tumours observed in a fatal context, the interpretation of the statistical analysis of occult tumours observed in an incidental context can be quite difficult. Incidental tumour data are analysed using methods which have been termed 'prevalence methods' (Hoel & Walburg, 1972; Peto, 1974; Peto *et al.*, 1980). With prevalence methods, the age range spanned by a carcinogenesis experiment is subdivided into discrete age intervals. For each experimental group of animals, a proportion is formed in each age interval as follows: the numerator is the number of animals that are found to have a tumour at necropsy after dying of unrelated causes during the interval, and the denominator is the total number of animals that die of causes unrelated to the tumour during the interval. The prevalence methods analysis tests for equality of the resulting proportions in control and exposed groups across all of the age intervals using standard contingency table methods. Although significant differences between control and exposed groups are taken as evidence that the test agent is associated with changes in the tumour onset rate (Peto *et al.*, 1980), the prevalence method can be shown to test for equality of the incidence rates of underlying tumours only under strict assumptions (McKnight & Crowley, 1984). Moreover, the proportions formed within each interval will estimate the true tumour prevalence rates only if animals dying of causes other than the tumour of interest are representative, with respect to tumour presence, of all animals surviving the interval (Lagakos & Ryan, 1985). Testing for equality of the proportions of animals dying with tumours is equivalent to testing for equal prevalence rates only if, within each age interval, the risk of dying for tumour-bearing animals relative to tumour-free animals is the same in control and exposed groups. If, for example, an exposed animal with a tumour is twice as likely as an exposed animal without a tumour to die during a particular age interval, then, for prevalence methods to be legitimately applied, a control animal with a tumour should also be twice as likely as a control animal without a tumour to die during that age interval.

If the relative risk of dying for tumour-bearing animals (relative to tumour-free animals) varies among control and exposed groups for some age intervals, then the prevalence methods may not provide valid tests for equality of tumour prevalence rates. The assumption that these relative risks are equal in control and exposed groups for all age intervals may not be realistic for all carcinogenesis experiments. Suppose, for example, that the presence of a tumour accelerates death in exposed animals more than in control animals for some cause unrelated to tumour. Then the relative risk of dying for tumour-bearing animals relative to tumour-free animals will tend to be higher in exposed groups than in the control group, at least in early age intervals. The added burden of even a nonfatal tumour in an animal which is exposed to a maximum tolerated dose of some chemical may put the animal at a substantially higher risk of death due to some cause not related to a tumour, while the same tumour in a healthy control animal would result in little increased risk. As in the related situation for fatal tumours (previously discussed), such an occurrence could give the appearance of tumour acceleration; however, the apparent acceleration would reflect differences in mortality patterns, not an acceleration of tumour incidence or tumour development. As with fatal tumours, incidental tumour analyses that indicate significant acceleration of tumour onset without an accompanying increase in the proportion of animals observed with tumours must be interpreted with care.

2.13 Combining analyses of fatal and incidental tumour data

For any particular organ, it may be that tumours are observed both in a fatal and an incidental context. If, for example, a group of animals has a high intercurrent mortality rate, many potentially fatal tumours may be observed in an incidental context because tumour-bearing animals die from other causes before being killed by the tumours. A method of combining the analyses based on fatal and incidental tumours is presented in Chapter 5. When the contexts of observation of all tumours are to be used in evaluating an animal carcinogenesis experiment, the determination of carcinogenic potential at a particular organ must be based on this combination of fatal and incidental tumour analyses (Peto *et al.*, 1980). Evaluation of carcinogenicity based either on fatal tumours separately or incidental tumours separately can be misleading. For example, consider an experiment in which the age-specific incidence rates for some type of tumour are equal in the control and exposed groups, and the proportion of these tumours which would eventually be observed in a fatal context is the same in the control and exposed groups. Suppose, however, that the exposed animals with these tumours have a higher risk of dying from causes not related to tumours than do the control animals with the same tumours. Then, many of the exposed animals with these tumours that would eventually have been observed in a fatal context may have their tumours observed in an incidental context because the intercurrent mortality rates have been selectively increased. Thus, if only the fatal tumour analysis is considered, there might appear to be a deficit of fatal tumours in the exposed groups. Such a selective increased risk of intercurrent mortality, however, would lead to a higher relative number of exposed animals observed with tumours in an incidental context. Thus, in the combined analysis, the deficit of exposed animals with fatal tumours would, in some sense, be balanced by an excess of exposed animals with incidental tumours. The combination of the two analyses, one based on tumour death rates and the other based ostensibly on tumour prevalence rates, may seem somewhat contrived and excessively complex. In fact, it is difficult to justify such an analysis rigorously (McKnight & Crowley, 1984). This analysis, however, is presently the best solution to the difficult problem of using information on the ages at death of tumour-bearing animals to test for differences in tumour incidence rates. This problem arises whenever the ages at observation of tumours are determined by the deaths of the tumour-bearing animals, because the ages at death may be influenced by factors unrelated to the tumours. Although animal sacrifice schemes can be formulated to alleviate the problem partially, it will be inherent in any experiment in which all animals are allowed to die a natural death (McKnight & Crowley, 1984).

In order to justify formally the use of the fatal and incidental tumour analyses to make inferences about tumour incidence rates, certain assumptions, some of which are difficult to verify, must be made. In particular, an assumption similar to that of noninformative censoring is required (Kodell *et al.*, 1982a; Lagakos, 1982). Such an assumption implies that, with regard to all life-shortening disorders (including toxicity) not caused by the presence of a tumour, a tumour-bearing and tumour-free animal are equally healthy. In other words, it is assumed that tumour-bearing and tumour-free animals are equally susceptible to, and have the same distribution for time to death due to, each life-shortening disorder. There is some experimental evidence to argue against such an assumption (Lagakos & Ryan, 1985), and there is additional indirect evidence which suggests it may not always be reasonable. The presence of a tumour has been shown to impair certain specific immune functions in rodents (Howell *et al.*, 1975; Nelson *et al.*, 1980; Perry & Greene, 1981); in addition, several carcinogens have been shown to be immunotoxic for a number of functions (Ball, 1970; Parmiani *et al.*, 1971; Vos & de Roij, 1972; Gainer & Pry, 1972; Koller, 1973). Thus, it is possible that tumour-bearing animals may be more susceptible than tumour-free animals to disease in some cases, and that this effect could be more pronounced in exposed groups than in control groups. Whether or not the assumption of noninformative censoring is reasonable can be determined only by experiments designed to compare the general health status of tumour-bearing and tumour-free animals. It should be noted, however, that the presence of immunotoxic effects such as those mentioned above may have little impact on mortality patterns in experiments in which test animals are housed under conditions free from pathogenic organisms.

2.14   Context of observation unavailable

In many studies conducted to date, the contexts of observation are unknown for all or most tumours. As definitive information in this regard will probably remain incomplete in many future studies, a method of adjusting for differential survival rates which does not require the accurate categorization of individual tumours as incidental or fatal would be useful. Although no such method is currently available, the lack of information regarding contexts of observation of tumours does not necessarily preclude a valid assessment of the carcinogenic potential of a test compound. In some cases, it may be possible to classify particular lesions as being almost always fatal or almost always incidental, although there is currently no consensus among pathologists on this point. Another possible approach is to carry out two separate analyses, one taking all tumours to be incidental and the second taking all tumours to be fatal. In many cases, these two analyses will be in agreement, lessening concern regarding the lack of data on context of observation. It should be cautioned, however, that the actual level of significance attained by an analysis using accurate information on contexts of observation may not always be bracketed by the levels of significance that are attained in the separate analyses performed by assuming that all tumours are incidental or all tumours are fatal (Lagakos & Louis, 1985). Moreover, doubling the number of statistical tests will accentuate the problem of multiple comparisons. In some cases, no valid assessment of carcinogenic potential can be made without knowing the context of observation (Peto et al., 1980). Therefore, attempts to acquire reliable information on context of observation are to be encouraged.

2.15 Analysis of crude tumour rates

In view of the difficulties involved in determining the context of observation of a tumour (that is, fatal or incidental) and in interpreting the subsequent analyses based on ages at death of tumour-bearing animals, it has been suggested that the analysis of crude tumour rates be performed, as a first step in evaluating tumour incidence data (Gart et al., 1979). A crude tumour rate for an experimental group is defined as the number of animals in the group which develop a tumour (regardless of the context of observation) at some organ during the experiment, divided by the number of animals in the group in which that organ was examined for the presence of a tumour. To avoid any possible bias due to differential early mortality, it is often advisable to eliminate all the animals that died in the experiment prior to the occurrence of the first tumour when crude tumour rates are computed (Gart et al., 1979). Although differences among groups with respect to longevity can seriously affect the analysis of crude tumour rates, one advantage of analysing the crude rates is that the impact of differences in mortality can readily be predicted. The assumption made in assessing the effect of mortality on crude rates is that the longer an animal survives, the greater is its probability of developing a tumour. Thus, if two groups of animals have identical (unobservable) age-specific tumour rates, but one group has higher intercurrent mortality rates, then the group with the higher intercurrent mortality rates will tend to have a lower crude tumour rate. The effect of differences in mortality upon the analysis of crude tumour rates is summarized in Table 2.2 (reproduced from Gart et al., 1979). Under the column for tumour association with treatment, ' + ' indicates a significant increase in crude tumour rates associated with increasing exposure level, '0' indicates no significant change in the crude tumour rates with increasing exposure level, and ' − ' indicates a significant decrease in crude tumour rates associated with increasing exposure level. Under the column for mortality association with treatment, ' + ' indicates increasing mortality with increasing exposure level, '0' indicates no change in mortality with increasing exposure level, and ' − ' indicates decreasing mortality with increasing exposure level.

The table indicates certain situations in which the analysis of crude tumour rates may suffice to provide evidence of the carcinogenicity of a test agent. In particular, when

Table 2.2 Interpretation of the unadjusted analyses of tumour incidence in light of the survival analyses

Outcome type	Tumour: association with treatment	Mortality: association with treatment	Interpretation[a] of the unadjusted test of tumour incidence
A	+	+	Unadjusted test may underestimate tumorigenicity of the treatment
B	+	0	Unadjusted test gives a valid picture of the tumorigenicity of the treatment
C	+	–	Tumours found in treated groups may reflect the longer survivorship of the treated groups. A time-adjusted analysis is indicated
D	–	+	The apparent negative finding in tumours may be due to the shorter survivorship in the treated groups. A time-adjusted analysis and/or a re-test at lower doses is indicated
E	–	0	Unadjusted test gives a valid picture of the possible tumour-preventive capacity of the treatment
F	–	–	Unadjusted test may underestimate the possible tumour-preventive capacity of the treatment
G	0	+	High mortality in treated groups may lead to the unadjusted test missing a possible tumorigen. Adjusted analysis and/or a re-test at lower doses is indicated
H	0	0	Unadjusted test gives a valid picture of the lack of association with treatment
I	0	–	Longer survivorship in treated groups may mask a tumour-preventive capacity of the treatment

[a] Many of these interpretations assume that the maximum tolerated dose (MTD) was used and that a sufficient proportion of animals survived in sufficient numbers for an appropriate length of time

testing for increases in tumour incidence rates in exposed animals, a definitive statement regarding carcinogenicity can usually be based on the analysis of crude rates in outcomes A, B, E, F, H, and I given in Table 2.2. In situations with outcome A in which the differences among crude rates are of marginal significance, adjusted analyses may be desired to account for the differential mortality, thus improving the strength of the statistical evidence regarding carcinogenicity. Adjusted analyses refer to the previously discussed analyses of fatal and incidental tumours using the ages at death of tumour-bearing animals. Adjusted analyses usually are essential only in outcomes C, D, and G. There are clearly situations in which analyses of crude rates do not allow a definitive statement regarding carcinogenicity. However, in cases in which the crude analyses do suffice, the interpretation of the results is straightforward, and no unverifiable assumption is needed to justify the method of analysis. In some situations (for example, if it is known that a type of tumour is rapidly lethal or if acceleration is suspected), both adjusted and unadjusted analyses may be valid, but the adjusted analyses can have greater power.

2.16 Concomitant information

Although the analysis of data on survival and tumour pathology is of primary importance in evaluating a carcinogenesis experiment, other data which are sometimes useful in evaluating carcinogenic potential may be obtained from long-term animal experiments. For example, both survival and tumour rates can vary with varying patterns of weight gain and food consumption (Weindruch & Walford, 1982; Haseman, 1983a). Thus, observations of concomitant variables, such as animal weights, food consumption and measurements from haematology or other clinical chemistry examinations, are often recorded at various times during the course of an experiment (IARC, 1980). Whenever possible, such observations should be recorded for individual animals and not merely summarized by cage or group average. By use of stratification, analysis of tumour incidence data can be adjusted quite easily for differences in certain concomitant variables, such as initial body weight. Alternatively, logistic regression methods can sometimes be useful in adjusting for concomitant variables (Dinse & Lagakos, 1983). Formal incorporation into the statistical analysis of those variables for which observations are obtained at various times during the course of an experiment can be technically difficult. Furthermore, interpretation of such analyses is often not straightforward. If, for example, treatment causes an observable response which is on the causal pathway to carcinogenesis, then adjusting for this observable response in the analysis of tumour rates can lead to incorrect inferences regarding carcinogenicity. Nevertheless, observations on such variables often provide valuable information necessary to the valid interpretation of a carcinogenesis experiment (see, for example, the discussion section in Tarone *et al.*, 1981).

2.17 Need for interdisciplinary decision process

A variety of statistical methods for the analysis of tumour pathology data from a long-term animal carcinogenesis experiment will be presented in this book. The methods which are appropriate for a given experiment are determined primarily by the extent of the pathology reporting (that is, whether or not the contexts of observation of all tumours are reported) and by the presence or absence of differences between the control and exposed groups with respect to intercurrent mortality. Whatever statistical methods are used, it should always be kept in mind that the statistical analysis is only one component in the evaluation of an animal carcinogenesis experiment. The process of carcinogenesis is extremely complex, and the proper evaluation of tumour pathology data requires the careful appraisal of intricate patterns of lesions, some of which are malignant and some benign. Certain complicated patterns of response may not be quantified easily using statistical methods. Thus, the evaluation of long-term carcinogenicity experiments must be an interdisciplinary process, incorporating the input of pathologists, toxicologists and other scientists.

2.18 Overall evaluation of carcinogenicity

For the overall evaluation of the carcinogenicity of a given exposure, say a chemical, many considerations are required. As far as data from long-term animal experiments are concerned, there may be several studies available, and, in addition, there will also be information on short-term tests and epidemiological studies. Thus, within the IARC programme on the Evaluation of the Carcinogenic Risk of Chemicals to Humans, the assessments of evidence for carcinogenicity from studies in experimental animals are classified into four categories (IARC, 1982b):

'(i) *Sufficient evidence* of carcinogenicity, which indicates that there is an increased incidence of malignant tumours: (a) in multiple species or strains; or (b) in multiple experiments (preferably with different routes of administration or using different dose levels); or (c) to an unusual degree with regard to incidence, site or type of tumour, or age at onset. Additional evidence may be provided by data on dose-response effects, as well as information from short-term tests or on chemical structure.

'(ii) *Limited evidence* of carcinogenicity, which means that the data suggest a carcinogenic effect but are limited because: (a) the studies involve a single species, strain, or experiment; or (b) the experiments are restricted by inadequate dosage levels, inadequate duration of exposure to the agent, inadequate period of follow-up, poor survival, too few animals, or inadequate reporting; or (c) the neoplasms produced often occur spontaneously and, in the past, have been difficult to classify as malignant by historical criteria alone (e.g., lung and liver tumours in mice).

'(iii) *Inadequate evidence,* which indicates that because of major qualitative or quantitative limitations, the studies cannot be interpreted as showing either the presence or absence of a carcinogenic effect; or that, within the limits of the tests used, the chemical is not carcinogenic. The number of negative studies is small since, in general, studies that show no effect are less likely to be published than those suggesting carcinogenicity.

'(iv) *No data* indicates that data were not available to the Working Group.

'The categories *sufficient evidence* and *limited evidence* refer only to the strength of the experimental evidence that these chemicals are carcinogenic and not to the extent of their carcinogenic activity nor to the mechanism involved. The classification of any chemical may change as new information becomes available.'

Similar criteria applicable to the epidemiological context are used in the assessment of evidence for carcinogenicity from studies in humans. The final evaluation of carcinogenic risk to humans relies strongly on the epidemiological information but also incorporates the evidence from short-term tests and from studies in experimental animals.

3. EXPERIMENTAL DESIGN

CHAPTER 3

EXPERIMENTAL DESIGN

3.1 Introduction

Although the primary emphasis in this monograph is on the analysis of car-cinogenicity data, several statistical principles underlie the design of all types of experiments. These need to be taken into account in the planning stages of any study, preferably with the involvement at that time of a qualified, experienced statistician. If these principles are ignored, sensible conclusions often cannot be drawn from the data, no matter how sophisticated the statistical method of analysis.

In addition to statistical principles, many other considerations are involved in the planning of a long-term carcinogenicity study. These include the responsibilities of key personnel involved in the conduct of the study, the characterization of the physical and chemical properties of the test substance, the selection of a suitable animal model, the control of the laboratory environment in which animals are housed, the route of exposure and the dosing regimen to be followed, health monitoring and procedures for both gross and histopathological examination, and the methods for accurately recording and storing data for subsequent statistical analysis. This chapter will focus primarily on statistical issues in the planning of experiments; however, this represents only one aspect of good design and cannot be considered in isolation from the many practical concerns just noted.

The need for a well-organized, overall experimental design is well stated in the introductory section of the IARC (1980) report on the conduct of carcinogenicity tests:

'The objective of any chronic study is to ascertain what effect repeated administra-tion of a chemical will have upon tissues or organ systems in animals of either sex of the test species. The attainment of the objective requires:

'(a) a well-devised and explicit protocol, coupled with sufficient supervision to monitor the daily activities of the study, to ensure that all items of protocol and any changes thereof are understood and are being followed. Any deviations from the protocol must be well documented as to the reason(s) for the deviations, their extent and their nature;

'(b) a technical staff who thoroughly understand their responsibilities and duties, as well as a management that recognizes the importance of the technical staff in the conduct of a carcinogenicity study and is supportive of the staff;

'(c) a record-keeping system which is accurate, reliable, secure and complete . . . ;

'(d) a health monitoring programme that will ensure accurate diagnosis of disease or toxic states, with a minimal loss of tissue samples for histological examination; and

'(e) an archive for storing the test data, protocols and specimens to allow for possible reevaluation in the light of future studies.'

Objectives of carcinogenicity testing

A long-term carcinogenicity bioassay may be conducted for one of several purposes; these may include (i) screening for potential carcinogens, (ii) determination of dose-response relationships, and (iii) elucidation of possible mechanisms of carcinogenic action (Clayson *et al.*, 1983). Screening studies may be employed in order to evaluate the carcinogenic potential of the compound on a qualitative basis. Such studies are intended to establish the existence of a carcinogenic hazard. They are usually conducted at relatively high exposure levels that are administered for the major portion of the lifespan of the test species, so that the probability of observing an effect in a relatively small sample of experimental animals is maximized. Although the observation of adverse effects at a single dose can provide positive evidence of carcinogenicity, a second, lower dose is frequently employed, both to confirm the evidence and to protect against the loss of the high-dose group to intercurrent mortality.

A series of increasing dose levels might be employed in a dose-response study intended to delineate in more quantitative terms the shape of the dose-response curve for a carcinogenic agent. This information is useful if some measure of the potency of the agent is to be estimated (Purchase, 1980), or if it is desired to extrapolate results from the high doses actually used down to lower doses more characteristic of the human environment (Crump, 1979; Armitage, 1982).

More elaborate experimental designs may be used in an attempt to define the mechanism of action of the test agent. However, as studies of transplacental carcinogenesis, initiation/promotion systems, variable exposure patterns (including cessation of exposure), and the synergistic/antagonistic effects of mixtures become more commonplace, guidelines need to be developed for the design of these more specialized experiments.

Experimental design

Complete specification of the experimental design requires careful consideration of a number of design parameters. Perhaps the most fundamental parameter is the total number of animals to be used. While the amount of information obtained will clearly increase with the use of additional animals, there comes a point where the value of the incremental information may not be worth the extra cost. (A more precise discussion of this point is given in Section 3.3 below.) From the practical point of view, logistical and budgetary constraints may also serve to limit the size of the experiment.

Another fundamental consideration is the number of treatment groups and the particular dosing regimen to be applied to each group. As was discussed in the previous section, a variety of possible treatments and dosing regimens may be considered, depending on the objectives of the study in question.

Once the size of the experiment and the structure of the experimental treatments have been determined, there remains the question of how to allocate the available animals to the various groups in the best possible way. The simplest approach is to assign equal numbers of animals to each group. While such a balanced allocation may be reasonable in many cases, it is not always the optimal strategy (see Sections 3.3–3.5).

A crucial part of any experimental design is proper randomization. Without a randomized design, it is not possible to determine whether any observed differences between the treatment groups are more likely to be due to the treatments themselves than to other intrinsic differences between the groups. Proper randomization permits statistical inferences based on the probability of any observed effects being due to chance alone. Thus, randomization both protects against biases due to unsuspected confounding factors present in the laboratory environment and provides a basis for valid statistical inference (Cochran, 1974; Kempthorne, 1977; Gart *et al.*, 1979; Edgington, 1980).

General design guidelines

A number of important factors must be taken into account in the early stages of planning a carcinogenicity bioassay. These have been considered by a variety of expert groups (Health and Welfare Canada, 1975; Food Safety Council, 1978; Interagency Regulatory Liaison Group, 1979; Environmental Protection Agency, 1979; IARC, 1980). Although many of these are nonstatistical in nature, they nonetheless represent important components of the overall experimental design.

Characterization of test substances: Prior to the conduct of the study, the physical and chemical properties of the test substance should be established, including its purity and the nature of any impurities. Knowledge of chemical reactivity is of importance, particularly when the test agent can react with components of the basal diet.

Species and strain of test animal: In theory, the ideal animal model would be one in which there is little or no tumour incidence in the control animals, but in which the effects of the treatment being tested for carcinogenicity are easily seen. It is clear that the choice of a strain highly resistant to tumours is a poor idea. However, it is not so obvious that a species which has a high background incidence of tumours of the same type as the ones being studied is also to be avoided. There are two main reasons for this. One is that, with low spontaneous rates, far fewer animals of a strain with low background incidence are needed to detect a small increase as statistically significant (see Section 3.3). The second reason for avoiding strains with high background incidence rates is that it is not clear whether an increase in the rate of occurrence of a tumour, common in the animal species, but less so in man, really provides biological evidence in support of a carcinogenic effect in man (Clayson *et al.*, 1983).

In practice, limitations of space, time and cost usually dictate the use of small rodents, particularly the rat, mouse and Syrian hamster, in carcinogenicity bioassays. The cost of the animals, as well as their availability, longevity and familiarity, are all important factors affecting the choice. Because of differences in susceptibility, the use of more than one species or strain can be advantageous.

Route of exposure: In order to facilitate interspecies conversion, it is desirable that the route of exposure correspond to that in the human situation. Alternative routes may be acceptable if they result in equivalent levels of the test material or its metabolites in the target tissue.

Duration of the experiment: The duration of the experiment can markedly affect the conclusions reached. In general, it is desirable to continue the experiment for a period sufficiently long to provide enough time for the development of tumours which occur long after the start of the exposure period. On the other hand, extending the duration of the experiment may result in a high rate of occurrence of spontaneous lesions among the control animals. For example, in the absence of treatment, more than 20% of female B6C3F1 mice develop lymphoma-leukaemia of the haematopoietic system after 104 weeks of age (Ward *et al.*, 1979) and over 90% of male F344 rats develop interstitial-cell tumours of the testis by the same time (Goodman *et al.*, 1979). The high rate of occurrence of these and other lesions in untreated aged animals makes it more difficult to identify significant effects in the exposed groups at such sites. The interpretation of data involving aged rodents may also be complicated by normal geriatric changes which occur within animal populations (International Life Sciences Institute, 1984a).

In the past, experiments of 18 and 24 months have often been used for mice and rats, respectively, with exposure beginning at the time of weaning. Two other criteria for the termination of the study have also been utilized. One is to continue the study for the full lifespan of the experimental animals. With modern animal husbandry, however, it is possible that some rodents may live for three to four years or even longer. Another proposal has been to terminate the experiment when the proportion of animals surviving in the control group falls to 20%, so that an appreciable number of rodents will be available for comparisons at terminal sacrifice. As discussed in Chapter 5, however, a proper evaluation of the experimental results will take into account all animals included in the experiment regardless of whether or not they survived to the end to the study. Adaptive terminal sacrifice methods have been studied by Louis and Orav (1985), but, at present, the proposed methods are not practicable.

Interim sacrifices: In many studies, a number of animals may be sacrificed at predetermined points in time during the course of the study. In a 24-month rat study, for example, ten animals in each treatment group may be sacrificed at both 12 and 18 months. This will facilitate study of the progressive pathogenesis of the lesion of interest and will ensure that both exposed and unexposed animals are available for purposes of comparison at these times. Efficient methods of planning adaptive interim sacrifices have been considered by Bergman and Turnbull (1983) and by Louis and Orav (1985), although these methods require rapid pathological examination.

Dose selection: Dose selection is one of the most controversial and important elements in the development of a protocol for a chronic bioassay. In addition, considerations behind the selection of appropriate dose levels depend to a large extent on the objectives of the study at hand.

In screening studies, the biological ideal would be to test only dose levels comparable to those to which humans are exposed. However, this is not practicable on statistical and economic grounds, unless the substance tested is an extremely potent carcinogen. There are potentially millions of humans exposed to many of the test substances under consideration, whereas it is usually feasible to expose only hundreds or perhaps a few thousand animals. Thus, a substance that causes the rate of some cancer in humans to increase from 1% to 2%, say, might cause tens of thousands of human cancers but might not be detected as a carcinogen even in a relatively large animal experiment. As it is not feasible to carry out tests involving millions of animals, the only solution is to use dose levels that induce measurable rates of response.

It is essential to have more than one dose level, for several reasons. One important purpose is to provide for the possibility that a misjudgement has occurred with the choice of a single high dose, resulting in either few animals surviving long enough for tumours to arise, or such severe toxic effects are seen that the relevance of the findings are doubtful. Secondly, a treatment effect that is dose-related over several levels of exposure is more convincing than one that is demonstrated only in a single-dose group. A third reason is to allow for the possibility that metabolic pathways used at a high dose may differ from those used for lower doses. If this consideration is ignored, a substance causing tumours at very high dose levels by a mechanism that does not occur at lower dose levels may be erroneously deemed unsafe for humans. A fourth reason is that it is reassuring if no large effect occurs at dose levels in the range to which humans are exposed.

The results of a recent survey of published guidelines of experimental designs for carcinogenicity tests are summarized in Table 3.1 (International Life Sciences Institute, 1984b). An examination of these guidelines indicates that a control and two or three positive levels have often been recommended. The highest dose should elicit minimal signs of toxicity to ensure that the test animals have been sufficiently challenged, yet not be so great as to result in appreciably decreased body weight or decreased survival, other than as a result of tumour induction. Lower doses are taken either to be specified fractions of the high dose or to lie within a certain range of the high dose.

The highest dose that satisfies the preceding criteria is often referred to as the 'maximum tolerated dose' or MTD (Munro, 1977). It is possible that the MTD could be different for males and females of the same species. If the difference between sexes is small, a common MTD can be employed. If the difference in sex-specific MTDs is appreciable, separate MTDs may be employed, at the expense of comparability between sexes, in order to maximize the sensitivity of statistical tests within each sex for increased tumour occurrence in the high-dose group relative to the control group.

While the guidelines given in Table 3.1 are generally quite specific, their underlying rationale is often less explicit. Most are vague with respect to recommendations concerning randomization. In this chapter, however, we will attempt to develop recommendations based on sound statistical principles of experimental design.

Table 3.1 Summary of selected recent guidelines on experimental design for carcinogenicity bioassays

Source	No. of dose levels (excluding untreated controls)	No. of animals of each sex per dose	High dose	Low dose	Intermediate doses
EPA[a] (1979)	3+	50	Induces slight toxicity but no substantial reduction in longevity due to effects other than tumours	Less than $\frac{1}{2}$ of intermediate doses but not less than $\frac{1}{10}$ of high dose	$\frac{1}{4}$ to $\frac{1}{2}$ of high dose
IRLG[b] (1979)	2+	No. of animals required to provide adequate assurance of safety if the test failed to detect carcinogenicity	Can be administered for the lifetime of the test animal and not (i) produce clinical signs of toxicity or pathological lesions other than those related to a neoplastic response, (ii) alter the normal longevity of the animals from toxic effects other than carcinogenesis, and (iii) appreciably inhibit normal weight gain		
IARC[c] (1980)	2	50	Elicits some toxicity when administered for the duration of the test period, but does not induce (i) overt toxicity, (ii) toxic manifestations which are predicted materially to reduce the life span of the animals except as the result of neoplastic development, or (iii) 10% or greater retardation of body weight gain as compared with control animals	$\frac{1}{4}$ or $\frac{1}{2}$ of high dose	
OECD[d] (1981)	3	50	Elicits signs of toxicity without substantially altering the normal lifespan due to effects other than tumours. For diet mixtures, the ingested concentration should not exceed 5%	Should not interfere with normal growth, development or longevity of the animal or result in any indication of toxicity. In general, not less than 10% of high dose	Mid-range between high and low doses depending upon the toxicokinetics of the chemical

[a] Environmental Protection Agency
[b] Interagency Regulatory Liaison Group
[c] International Agency for Research on Cancer
[d] Organization for Economic and Cooperative Development

Chapter overview

In Section 3.2, the basic principles of experimental design, including randomization, replication and stratification, are reviewed. Practical considerations involving the design and conduct of carcinogenicity trials are also noted. Experimental designs for screening studies are considered in Section 3.3. In addition to guidelines on sample size requirements for conventional bioassays, consideration is given to experiments involving multiple strains of test animals. Dose-response studies are discussed briefly in Section 3.4; special studies designed to elucidate certain aspects of carcinogenic mechanisms of action are treated in Section 3.5. Design considerations relating to histopathological analysis are considered in Section 3.6, while operational procedures involved in the acquisition and recording of experimental data are discussed in Section 3.7. A brief summary of the recommendations made in this chapter is given in Section 3.8.

3.2 Principles of experimental design

The primary purpose of experimental design is to ensure that the objectives of the study can be met and that valid, meaningful conclusions can be drawn from the results obtained. A good experimental design will also maximize the value of the information obtained by eliminating potential sources of bias, reducing experimental error to a minimum, and providing means of assessing experimental error. This is accomplished through proper application of the techniques of randomization, replication and stratification.

Experimental units

The term 'experimental unit' refers to the smallest unit of experimental material which is treated alike. In some studies, for example, several animals are housed in the same cage. It is thus possible that interactions between animals in the same cage, in a common environment, could result in cage effects. In this event, the cage rather than the individual animal would constitute the experimental unit.

The experimental evidence for or against the existence of cage effects is scant, because such effects have gone largely unassessed in past studies. In one large skin-painting study in which such effects were considered, a highly significant ($p < 0.01$) indication of cage effects was found among two of the three positive-control groups (Gart, 1976, p. 113). However, significant effects ($0.01 < p < 0.05$) were found in only three of 52 tobacco-condensate groups. No cage effect was noted with respect to neoplastic lesions in a bioassay of hexachlorobenzene conducted by Arnold *et al.* (1985).

While it avoids cage effects, individual housing may lead to increased stress in rodents (Hatch *et al.*, 1965; Sigg *et al.*, 1966), although conflicting evidence is available in this regard (Andervont, 1944; Fare, 1965). If multiple housing is used, on the other hand, feed and water consumption must be administered collectively, and animals may be lost to cannibalism or the more rapid spread of communicable disease. In order to clarify these issues, some basic research on the desirability of group housing and the attendant possibility of cage effects is required from the point of view of proper experimental design.

The presence of other effects may also lead to an experimental unit which does not correspond to the individual animal. In two-generation studies involving exposure *in utero,* for example, animals from the same litter share common genealogical traits and are subjected to similar levels of transplacental exposure. As a result, the entire litter rather than the individual pup may constitute the experimental unit. Another example is that in which an entire column or contiguous group of cages is subjected to the same treatment. If there were a gradient within the laboratory influencing tumour occurrence, the ensemble of cages could conceivably represent the experimental unit. Such effects may be due to differences in susceptibility among the test animals, or to differences in environmental factors such as temperature, lighting, humidity and air flow (Fox *et al.,* 1979).

In a review of data from experiments on the carcinogenic potential of the food-colour additive, FD & C Red No. 40, Lagakos and Mosteller (1981) noted a correlation between the incidence of reticuloendothelial tumours and the animal's position on the shelf-level on racks, which persisted after adjustment for sex, dose and rack column. The tumour rates appeared to be higher in animals on the upper shelves. Animals in successive groups were allocated to successive rack positions. In this particular study, group 1 males went into the top three (of five) shelves of the front of rack 1, group 2 females went into the bottom two shelves of the front of rack 1 and the top shelf of the back of rack 1, and so on. Thus, shelf position introduced some bias into the treatment comparisons, which did not, however, affect the overall negative conclusions. Another example of potential positional effects is the study that has been reported by Greenman *et al.* (1984).

As our knowledge of clustering effects is still somewhat limited, the analysis of most studies continues to treat the individual animal as the experimental unit. This will of course be valid when the experimental design is such that the individual animal is the appropriate unit for purposes of statistical analysis, as in the completely randomized design with individual housing that is discussed in the next section. This may also be reasonable when any cage, litter or positional effects that may be present are negligible. However, empirical evidence to support this assumption would be reassuring.

Randomization of animals

As noted earlier, the purpose of randomization is two-fold. First, it ensures against potential biases in the experimental results. Second, valid statistical inference can be based on the permutation distribution induced by the randomization scheme employed. This avoids the need to make further assumptions concerning underlying statistical models for the experimental data. In order to avoid bias, it is essential that the predisposition to the response of interest be the same in all treatment groups. Bias will be introduced if the animals in one group are more likely to develop tumours than those in another group.

To avoid bias, animals must be assigned randomly to treatment groups or cages. This is done usually by using so-called 'pseudo-random' numbers generated by a computer. Depending on the number of animals to be used, the number of animals caged together, and the number of experimental groups, a randomization list may be drawn up.

As an example, we consider the situation of 200 animals of the same sex to be assigned randomly to four groups of 50 with five animals from the same group caged together. The 200 animals, thus, have to be assigned to 40 cages and these to the four treatment groups. As a preliminary step, consecutive numbers should be arbitrarily given to all 200 animals; this may well be their order of presentation.

The randomization list will be a random sample without replacement from the 40 cage numbers, each being included exactly five times. For example, the first ten numbers in such a list may be:

Consecutive animal number:	1	2	3	4	5	6	7	8	9	10	...
Random cage number:	37	32	12	8	9	32	17	19	23	16	...

This would mean that animal No. 1 would be placed in cage No. 37, and animal No. 2 in cage No. 32, and so on. (Note that animal No. 6 will also be assigned to cage No. 32.) This procedure ensures that cage mates have been randomly selected.

This complete randomization of animals into cages permits distribution of these cages in a deterministic way to the treatment groups. For example, the first ten cages are assigned to group No. 1, the second ten to group No. 2, and so on. However, a randomization list could once again be used, taking a random sample without replacement from the four group numbers, and using each number exactly ten times. This list, when matched to the consecutive cage numbers, would identify the group that was allocated to each cage.

Random location of cages

Randomization of the animals to cages and into experimental groups does not yet ensure that the predisposition to the response of interest is the same in all treatment groups. Care must be taken also that potential treatment effects are not confounded with environmental factors. An example of a design with potential for serious bias is shown in Figure 3.1 (Bickis & Krewski, 1985). Even if the animals have been randomized to cages, the existence of an environmental gradient, say, from left to right, would bias any comparisons between the various treatment groups. In particular, if tumour occurrence were enhanced along this gradient and an agent with no carcinogenic potential whatsoever were administered to the test animals, then the increasing trend in tumour incidence going from left to right would give the illusion of a dose-related effect.

A somewhat better design that has been used often in the past is the systematic cyclic design (Figure 3.2A), in which the doses are assigned to the cages in rotation.

Fig. 3.1   Cage layout for a systematic block design (after Bickis & Krewski, 1985)

☐ Control   ☐ Low dose   ▨ Mid dose   ▨ High dose

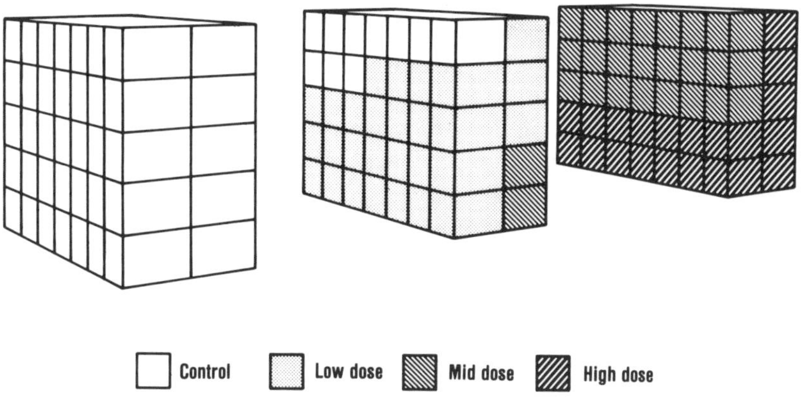

Although large biases may be avoided in this way, the use of a fixed sequence is potentially vulnerable to small biases. Another approach would be to assign the treatments to cages completely at random (Figure 3.2B). This procedure should provide adequate protection against environmental gradients with a moderate number of cages, and it has an element of simplicity that is readily exploited at the analysis stage (see Chapter 5). An objection to complete randomization that has been raised is that it increases the chance of misapplication of treatments; this problem may be controlled through careful record-keeping and the use of well-trained personnel. Visual devices, such as different coloured labels for the cages of different treatment groups, can be very helpful in this respect, although it does not allow for 'blind' delivery of treatment.

A more serious problem may be the potential for cross-contamination with volatile agents or through spillage of feed. In this case, the clustered block design in Figure 3.2C, in which all cages in the same column are treated identically, may be considered. With this last design, however, the column rather than the cage may form the most appropriate experimental unit.

Fig. 3.2   Cage layout for three experimental designs (front view of first bank of cages) (after Bickis & Krewski, 1985)

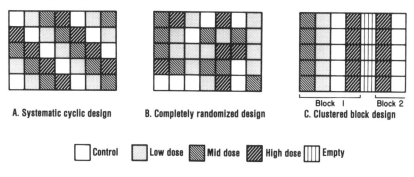

A. Systematic cyclic design   B. Completely randomized design   C. Clustered block design

Block I   Block 2

☐ Control   ☐ Low dose   ▨ Mid dose   ▨ High dose   ⦀ Empty

More complicated, Latin-square designs have been proposed by Lagakos and Mosteller (1981) as a means of balancing positional effects. However, their implementation requires that certain relationships be satisfied between the number of rows and columns in the cage rack and the number of dose groups. The symmetry which eliminates positional effects, moreover, would be disrupted easily by missing observations that cannot be avoided in long-term studies. Technically, such designs also require certain assumptions concerning the lack of interactions between rows, columns and treatments.

Another method of balancing positional effects and providing a more uniform environment is to rotate the cage positions during the course of the experiment. Although this procedure is sometimes used, it has not been subjected to the same careful study as the other designs. Nonetheless, some form of cage rotation may prove useful in designs with a high degree of clustering, such as the clustered block design. In this case, rotation of the columns would tend to minimize the effects of horizontal gradients. Similarly, frequent rotation of the positions of the three banks of cages in the systematic design shown in Figure 3.1 would serve to reduce the potential for serious biases.

While not necessary for purposes of bias reduction, the rotation of cages in a completely randomized design ensures a more uniform environment for all animals. By reducing this source of variation, it is possible that the sensitivity of any comparisons between treatments could be improved. For the same reason, consideration could be given to the periodic rotation of the cage positions within columns in the clustered block design or within banks in the systematic design. Until the contribution of environmental factors to the overall experimental error is more clearly defined, however, the potential gains from cage rotation remain unclear.

In any cage rotation scheme, some cages will become vacant due to deaths prior to the termination of the study. While the manner in which empty cages should be handled has received little attention so far, the inclusion of these cages in subsequent rotations will maintain the advantages of the original plan in terms of counterbalancing positional effects.

All of the preceding randomization schemes for cages are directly applicable when only one sex of the test animal is to be used. With protocols involving both sexes, there may be problems in housing males and females in adjacent cages. Because of this, it has been a common practice in the past to keep male and female animals apart, often in separate rooms. When this is done, any comparisons between sexes may reflect environmental as well as sex differences. With certain designs, however, it may be feasible to mix males and females to a limited extent. With the completely systematic design, for example, banks of males could be interspersed among banks of females. This would be advantageous when comparisons between sexes are of interest.

Regardless of the experimental lay-out and randomization scheme actually employed, care should be taken to document the precise placement of each cage and its corresponding treatment. Any subsequent alteration to this initial configuration should also be recorded meticulously. This information is essential in order to characterize the experimental design and to permit a valid statistical analysis of the experimental data.

Replication

The significance of any experimental finding will be greatly enhanced if the results can be reproduced. In order to assess reproducibility, some form of replication is necessary. Replication requires that several independent experimental units be subjected to precisely the same experimental conditions in order to provide an estimate of experimental error. This allows the statistical significance of any observed differences between treatment groups to be determined by comparing the variations between units in the same group.

In the completely randomized design with single caging, the individual animal is the experimental unit, and replication is achieved by increasing the number of animals allocated to each treatment. More information may be obtained if some grouping of subjects into homogeneous strata or blocks is done prior to the allocation (see section below). In this case, animals are assigned deliberately to the blocks to reduce variation, and then the animals within each block are assigned randomly to the treatments.

Another form of replication that has been used frequently in agricultural research, although not in carcinogenicity testing, is to repeat the entire experiment. In order that the individual experiments be comparable, they should all follow the same protocol. Aside from this constraint, however, there is no need to keep the repetitions uniform. It may, in fact, be advantageous to make them as diverse as possible. They could, for example, be carried out in different laboratories at different times and, if possible, use different suppliers of animals and feed and different batches of the test compound. If studies are replicated in this manner, it is possible to assess treatment-replicate interaction or the extent to which treatment effects may differ between replicates. The absence of such interaction provides additional assurance of the reproducibility of the result.

The use of replicates also has the advantage that the chance of discovering a susceptible subpopulation is increased (Haseman & Hoel, 1979). It is possible that the toxic effects of a compound are manifested only if certain other factors are present. These may be a genetic predisposition to toxic effects, the presence of certain other compounds in the diet, or, possibly, microbial or environmental influences. If only one replicate is used, then the chance of encountering the required set of conditions is less than if a number of replicates are employed under different conditions.

The demand for replication of experimental findings is inherent in the process of scientific evaluation. For example, in the *IARC Monographs* (IARC, 1982b), sufficient evidence for carcinogenicity in experimental animals is established only if 'there is an increased incidence of malignant tumours: (a) in multiple species or strains; or (b) in multiple experiments (preferably with different routes of administration or using different dose levels); or (c) to an unusual degree with regard to incidence, site or type of tumour, or age at onset.' Evidence based on replication across sexes and species is stronger than that based on experiments conducted under identical conditions.

Stratification

Many different factors, both genetic and environmental, can influence the process of tumour induction. For comparisons between groups to be as precise as possible, the animals in different treatment groups and their environments must be as similar as possible, other than with respect to the treatment of interest. If factors affecting tumour occurrence can be identified before the conduct of the study, the test animals can be divided into more homogeneous strata, defined in terms of different levels of such blocking factors. Comparisons between treatments may then be made within blocks or strata, thereby eliminating interblock variation from the comparisons of interest.

Animals can be stratified by litter so that genetic variation is reduced within a block (Mantel *et al.*, 1977; Mantel & Ciminera, 1979). In some studies, animals have been divided into different weight classes before being assigned to different treatments, thereby stratifying by initial body weight. If environmental effects in the laboratory are of concern, cage position might be used to define strata with, for example, one bank of cages constituting a block. Other variable factors that can influence the tumorigenic process include the age of the animals at the start of the study and the source of the animals, including the particular shipment from the supplier.

While stratification represents a potentially useful tool for increasing analytical precision, it is not without its disadvantages. Since blocks are different by design, the variation between blocks provides no information on experimental error. If the test subjects are few, stratification can actually diminish the amount of information on the magnitude of the experimental error even if it actually reduces the error. In the extreme case in which no two animals in the same block are treated alike, experimental error is estimated from the inconsistency of treatment differences across the blocks, or treatment-block interaction. If effects of treatment do differ among blocks, then this inconsistency can lead to an overestimation of experimental error.

Control animals

Regardless of the objective of the study, some form of reference or control group, against which the effects of the treatment of interest can be judged, is essential. The nature of the control group may, however, differ depending on the study protocol. For screening studies involving exposure to one or more levels of the test agent, the appropriate control is simply a group of unexposed animals. For more elaborate mechanistic studies, one or more different types of control group involving different treatments may be required in order to isolate the effects of interest (see Section 3.5).

It is essential that the treated groups differ from the control group only with respect to the treatment of interest, and not with respect to any other aspect, such as diet, husbandry or observation. Any comparisons between the treated and control groups will reflect all differences between these two groups. Thus, in an experiment with two groups – one in which animals are exposed to cigarette smoke in smoking machines and the other an untreated control group – the only assessment that can be made is the effect of smoking combined with the stress due to the the animals' being placed in the machines. To test the effect of smoking only, one needs control animals that are 'sham-smoked', that is, they are placed in machines for the same length of time as the treatment animals but not exposed to smoke. Similarly matched control animals will also be required when the treatment is applied by injection, with oral dosing by gavage or in a vehicle such as corn oil. Thus, if a chemical is administered in corn oil, the control animals should be administered corn oil without the chemical. Such control animals are termed 'vehicle control' animals to indicate that they have been subjected, as nearly as possible, to the method and route (i.e., the vehicle) of exposure experienced by the treated animals.

The question whether to include a positive control group, involving a known carcinogen, has been discussed extensively. However, it is generally agreed that in routine screening no positive control animals need be used (IARC, 1980). One reason is that carcinogens act by different mechanisms on different tissues, so that one would not necessarily know which positive control substance to choose when testing an agent of unknown carcinogenic potential. Furthermore, the inclusion of positive control groups introduces hazards for personnel and the risk of cross-contamination. However, when the objective is to assess the relative carcinogenicity of a range of treatments known to be carcinogenic (for example, cigarette-smoke condensate) and when the testing must be carried out in a series of studies, positive control animals are required, as it is otherwise impossible to compare reliably the carcinogenic potency of different treatments.

There has been some discussion about the use of two identical control groups (Society for Toxicology, 1982). With a completely randomized design, any difference between the two groups would be due solely to chance, so that, in effect, the two groups form one large control group. From the statistical point of view, differences between the two control groups could indicate systematic departures from complete randomization. Thus, while two control groups serve no useful purpose in a properly randomized experiment (other than to increase the total number of control animals), this practice could act as a quality control mechanism in terms of identifying unsuspected biases in design.

Haseman (1985) compared the response rates in the dual control groups used in a series of 16 bioassays of food-colour additives and found no significant difference between the response rates in the pairs of control groups used within the same study.

Criteria for evaluating experimental designs

The most important requirement of any experiment is that it should provide the imformation needed to meet the study objectives. In particular, it should be free from biases which may exist in the absence of randomization, and there should be a sufficient number of treatment groups to enable identification of the quantities of interest. Since there are generally many designs that satisfy these conditions, considerations of sensitivity and efficiency can be used to choose among them.

The sensitivity of an experiment is its ability to detect small differences. In screening studies, sensitivity is often quantified in terms of the false-negative rate, or the probability of not detecting a carcinogen of a given potency (see Section 3.3). Equivalently, sensitivity may be expressed in terms of the power, that is, the probability of detecting a carcinogen of a given potency. The expressions are equivalent because power equals one minus the false-negative rate. For dose-response studies, where the objective is usually the estimation of some parameter of the dose-response curve, sensitivity may be measured by the standard error of the estimate.

In general, experiments with more animals tend to be more sensitive. There may exist one design that will be the most sensitive among all designs of a given size. However, such optimal designs are not always practical, for the following reasons. First, the optimal design may depend on parameters that cannot be determined until the experiment is completed. Secondly, an experiment may have several objectives, and the design that is optimal for one objective may not be optimal for another. Finally, the optimal design may not be feasible because of operational or other constraints. Nonetheless, the optimal design may still be used as a yardstick for gauging the efficiency of the actual design relative to the optimal one.

3.3 Designs for screening studies

Conventional studies

In a screening study, the purpose of the experiment is to arrive at a decision regarding the carcinogenicity of the test compound. Two types of errors are possible in making such a decision: an innocuous chemical may falsely be declared carcinogenic (Type-I error), or a carcinogen may incorrectly be considered harmless (Type-II error). The probabilities of these two types of errors occurring are termed the 'false-positive' and 'false-negative' rates, respectively (Table 3.2). These error rates depend on both the experimental design and the decision procedure used. Once the experimenter decides on a false-positive rate, that is, determines the risk he is prepared to accept for making the first type of error, the decision rule will be derived from this mathematically. The predetermined value of the false-positive rate is termed the 'nominal significance level'. It is then the function of the experimental design to minimize the false-negative rate.

In order to gain some idea of the sensitivity of a screening bioassay, consider a simple experiment in which $n = 50$ animals are assigned to both a control and a single test group. Suppose that there is no difference between the groups with respect to intercurrent mortality and, thus, that the proportions of animals with tumours in the two groups are compared using Fisher's exact test at a nominal significance level of

Table 3.2 False-positive and false-negative rates in carcinogenicity screening tests

Experimental evidence for carcinogenicity	Carcinogen	
	No	Yes
No	Correct decision	False negative
Yes	False positive	Correct decision

Table 3.3 False-negative rates for a simple carcinogenicity screening test[a]

Excess tumour incidence in test group (%)[b]	Tumour incidence in control group (%)				
	0	1	5	10	20
5	90	88	87	88	90
10	43	49	61	69	77
15	11	18	34	46	58
20	2	5	15	25	36
25	<1	1	5	11	19

[a] Based on Fisher's exact test ($\alpha = 0.05$) with 50 animals in each of a control and a test group and assuming that all animals respond independently
[b] Difference between the response rates in the test and the control groups, respectively

$\alpha = 0.05$. (For details of this statistical test, see Chapter 5.) As indicated in Table 3.3, the false-negative rate for compounds inducing an increase of 25% over the background incidence rate is less than 1% whenever the spontaneous response rate is low. These results also suggest that a carcinogenic compound tested at a dose level inducing only a 5–10% increase over the background incidence rate might well go undetected. However, the use of high doses tends to maximize the carcinogenic potential of the test compound and thereby minimize the risk of a false-negative result.

Similar results for group sizes of $n = 25$, 50, 75 and 100 are shown in Table 3.4. When the background incidence rate is low, the use of more than 100 animals per group will result in moderate false-negative rates, with compounds inducing tumours in as few as 10% of the exposed animals, but at nearly double the cost. The use of 25 animals per group would be effective only for compounds responsible for an increased risk well in excess of 25% in exposed animals.

Minimum sample sizes required to detect a carcinogenic effect of a given magnitude with Fisher's test procedure may be calculated by summing the probabilities of those outcomes that would result in a significant result. This approach has been used by Haseman (1978) to obtain sample sizes for select values of p_0 and p_1, the response probabilities in the unexposed and exposed groups, respectively.

A simple approximation to the minimum value of n required to achieve specified error rates for two given response probabilities, p_0 and $p_1 = p_0 + \delta$, has been developed by Walters (1979). (A detailed comparison of this and other approximations has been made by Chen, 1984.) This particular result is based on the standardized difference

$$\Delta = \sqrt{2n}[\sin^{-1}\sqrt{p_0 - (2n)^{-1}} - \sin^{-1}\sqrt{p_1 + (2n)^{-1}}]$$

between the two proportions following the application of a continuity correction and an arc sine transformation. In testing at a nominal α level of significance, the power $1 - \beta$ of the Fisher test will be approximately

$$1 - \beta = (2\pi)^{-\frac{1}{2}} \int_{z_\alpha}^{\infty} \exp\left\{-\frac{(x - \Delta)^2}{2}\right\} dx,$$

Table 3.4   False-negative rates for a simple carcinogenicity screening test[a]

Excess tumour incidence in test group (%)[b]	No. of animals per group (n)	Tumour incidence in control group (%)				
		0	1	5	10	20
10	100	2	10	29	43	56
	75	12	21	41	55	66
	50	43	49	61	69	77
	25	90	89	85	84	87
15	100	<1	1	6	15	27
	75	1	3	15	27	41
	50	11	18	34	46	58
	25	68	67	67	71	77
20	100	<1	<1	1	3	9
	75	<1	<1	4	9	19
	50	2	5	15	25	36
	25	42	42	48	55	65
25	100	<1	<1	<1	<1	2
	75	<1	<1	1	2	7
	50	<1	1	5	11	19
	25	21	23	31	40	51

[a] Based on Fisher's exact test ($\alpha = 0.05$) with n animals in each of a control and a test group and assuming that all animals respond independently
[b] Difference between the response rates in the test and the control groups, respectively

where z_α denotes the $100(1 - \alpha)$ percentile of the standard normal distribution. By iterating on n, the minimum size required to achieve power $1 - \beta$ can be readily evaluated, given p_0 and p_1. This approximate procedure is computationally simpler than the direct approach, yet it yields results in excellent agreement with the exact results (Walters, 1979), as does the related closed-form expression of Dobson and Gebski (1986).

The minimum sample sizes required to achieve a false-negative rate of $\beta = 0.10$, using a nominal significance level of $\alpha = 0.05$ based on this procedure, are shown in Table 3.5 for selected values of p_0 and p_1. These results indicate that, when the background incidence rate is low, the use of 50–60 animals will permit the detection of effects involving about 15% of the exposed animals, subject to the specified error rates α and β. More animals would be required to detect a smaller effect or the same effect in the presence of a higher spontaneous response rate.

False-positive and false-negative rates may also be calculated while testing for increasing linear trends in proportion (see Chapter 5). Tests for linear trend may be based on large-sample chi-square statistics (Armitage, 1971, pp. 363–365) or on exact permutation tests (Cox, 1958; Thomas et al., 1977). Chapman and Nam (1968) obtained an explicit form for the asymptotic power of the former test, and Nam (1984) provides an expression for the exact unconditional power of the latter test.

Because the computations required for these exact results are extensive, some

Table 3.5 Minimum group sizes required to ensure a false-negative rate of 10% or less[a]

Excess tumour incidence in test group (%)[b]	Tumour incidence in control group (%)				
	0	1	5	10	20
1	819	2661	9084	16 287	28 110
5	162	243	503	783	1 232
10	80	100	166	233	339
15	53	61	90	119	163
20	39	44	59	75	98
25	31	34	43	53	67

[a] Based on Fisher's exact test ($\alpha = 0.05$) with n animals in each of a control and a test group and assuming that all animals respond independently
[b] Difference between the response rates in the test and the control groups, respectively

simpler, approximate results are desirable. Nam (1984) has derived a modified formula for sample size determination when testing for linear trend, based on a normal approximation with a continuity correction. For the special case of three equally spaced doses (including the control at dose zero) with n animals per dose, the minimum value of n required to result in Type-I and Type-II errors of α and β, respectively, is given by

$$n = A[1 + \{1 + [2(p_2 - p_0)/A]\}^{\frac{1}{2}}]^2/[4(p_2 - p_0)^2].$$

Here, p_0 and p_2 denote the response probabilities for the control and high-dose groups, respectively (p_1 is not involved in this term because of the equal-dose spacing) and

$$A = [z_\alpha (2\bar{p}\bar{q})^{\frac{1}{2}} + z_\beta (p_0 q_0 + p_2 q_2)^{\frac{1}{2}}]^2,$$

where $\bar{p} = \sum_{i=0}^{2} p_i/3$, $\bar{q} = 1 - \bar{p}$, and z_α denotes the $100(1 - \alpha)$ percentile of the standard normal distribution.

The number of animals, n, needed in each group to obtain 90% power with three equally spaced doses is given in Table 3.6. These results indicate that experiments with 50–100 animals per group will be effective in detecting a linear trend involving an increase of about 20% or more above the background incidence rate in the high-dose group. As with Fisher's exact test discussed above for pairwise comparisons, note that smaller sample sizes are required when the background rate is low.

Two-generation studies

One of the major considerations in the design of a two-generation bioassay is the selection of the second-generation animals. Studies on transplacental exposure using saccharin, styrene and amaranth have revealed considerable intralitter correlation (Grice et al., 1981). Although actual bioassay data are required to determine whether or not litter effects are to be expected also for tumours in second-generation animals, it seems clear that the litter rather than the individual pup should be considered as the experimental unit for the purposes of statistical analysis (see Section 7.6).

Table 3.6   Number of animals per group required to obtain false-positive rates of 5% and false-negative rates of 10% based on tests for linear trend with three equally spaced doses

Tumour response rates			Number of animals per group
p_0 (control)	p_1 (low dose)	p_2 (high dose)	
0.02	0.04	0.06	420
0.02	0.07	0.12	112
0.02	0.12	0.22	44
0.10	0.12	0.14	1150
0.10	0.15	0.20	224
0.10	0.20	0.30	70
0.20	0.22	0.24	1860
0.20	0.25	0.30	328
0.20	0.30	0.40	93

In the presence of appreciable litter effects, the statistical power of a two-generation study will depend on the number of pups selected from each litter. In order to illustrate the effects of intralitter correlation on statistical sensitivity, consider a simple hypothetical experiment in which 48 animals of the same sex are to be selected from the second generation in both a control and a single test group. The required number of animals could be obtained on the basis of one per litter, or two per litter from 24 litters. Fewer than 24 litters would be required if more than two pups per litter were chosen.

To illustrate the power of such a study, the probability of detecting a carcinogenic compound that induces a tumour incidence rate of 50% at a site where the background rate is 10% is shown in Table 3.7 for selected values of the intralitter correlation ceofficient in the test group (Krewski *et al.,* 1984a). These results reveal that, for a fixed sample size (in this case, 48), the statistical sensitivity is reduced with increasing intralitter correlation when more than one pup is selected from each litter. The goal of maximum sensitivity may thus be achieved by selecting only one pup per litter.

Table 3.7   Probability of significance in a hypothetical two-generation bioassay in the presence of intralitter correlation

No. of pups selected	No. of litters	Probability of significance (%)		
		Intralitter correlation in test group		
		0.1	0.5	0.9
1	48	95	95	95
2	24	93	88	83
3	16	90	79	68
4	12	87	68	49

For reasons of economy, Mantel (1980) has suggested that two or three pups be selected from each litter. However, since the cost of a two-generation study is due primarily to maintaining the second-generation animals for a period of about two years and to the associated histopathological diagnoses, the savings involved in breeding fewer litters over a 100-day period in the first generation may represent only a small fraction of the total cost of the study.

Multistrain studies

Another consideration in screening studies is the use of several strains of the test species (Haseman & Hoel, 1979). In order to assess the relative merits of single-strain and multiple-strain experiments, consider the following simplified genetic model. Suppose that a certain stock of animals consists of ten homogeneous subgroups of equal size, each with a spontaneous tumour incidence rate of 5%. Suppose further that a certain chemical will increase the tumour incidence rate to 25% in a certain number of these subgroups, but that it will have no effect on the other subgroups. In this model, the entire stock is intended to represent a single outbred strain, while each of the subgroups is either highly resistant or highly susceptible to the chemical under test, with the entire stock less sensitive overall than the susceptible subgroups.

Suppose now that an experiment is to be conducted in which 150 animals are to be assigned to a test group and an additional 150 animals to a control group. One strategy would be to choose these animals from the entire stock available. Alternatively, one or more of the ten subgroups might be selected at random, and a separate experiment conducted for each of the subgroups chosen. If two subgroups were selected, for example, two experiments with 75 animals in each of a treated and a control group would be involved. The first strategy corresponds to the case of a single outbred strain while the second strategy corresponds to the case of several inbred strains, with the total number of animals fixed at 150 in both cases.

The probabilities of detecting the carcinogenic potential of the chemical under test with these alternative strategies are shown in Table 3.8. These probabilities represent

Table 3.8 Probability (%) of detecting a carcinogen in multistrain experiments of equal size, assuming strains are chosen randomly from ten different strains

No. of strains	No. of animals per strain in each test group	No. of susceptible strains[a]			
		0	1	2	3
0[b]	150	3	12	30	53
2	75	4	23	39	54
3	50	3	28	48	64
5	30	2	30	51	66
6	25	2	32	54	70
10	15	3	32	53	67

[a] Tumour incidence increased from 5% to 25%
[b] Single outbred strain

the chances of observing a statistically significant increase in tumour incidence in any of the strains tested. When none of the ten subgroups is susceptible, the probabilities shown represent the chances of a false-positive result. Even though the multistrain experiments involve one statistical test for each strain, the overall false-positive rates are comparable regardless of the number of strains tested.

In all cases considered, the power of the strategy using multiple inbred strains exceeds that of the strategy which uses a single outbred strain. When two susceptible subgroups are present in the population, for example, the chances of detecting the carcinogen by using a single outbred strain are about 30%. With a multistrain experiment involving three randomly selected inbred strains, however, the chances of detecting the carcinogen are increased to about 48%.

The assumptions involved in the preceding example are no doubt over-simplifications. In practice, it is unlikely that the inbred strains used would be selected at random. Moreover, the model of the outbred population as a mixture of homozygous subpopulations does not provide for the heterozygosity present in actual outbred stocks.

While the choice of the incidence rates of the spontaneous and induced tumours is not critical, the use of a large number of inbred strains in a multistrain experiment will result in an insufficient number of animals per strain on which to base a meaningful analysis. If, for example, the design was constrained to 50 rather than 150 animals, the use of ten strains would result in only five animals per group. Another concern of a more practical nature is the difficulty of finding ten unrelated strains. Thus, while carcinogenic effects might be detected more readily by selecting several inbred strains for study when susceptible subpopulations do exist, some experience with this approach is required before it may be recommended as standard practice.

Sequential designs

In any carcinogenesis screening programme of a large pool of chemicals using animal experiments, some compounds will induce large tumour increases and can thus be classified easily as carcinogens, some compounds will induce no tumour increase of note and can thus be classified easily as noncarcinogens, but other compounds will give equivocal results which lead to no clear-cut classification. In the last case, decisions must often be made on the basis of the equivocal results, even when further testing would be advisable. Multistage experimental designs have been proposed to alleviate the problem of making decisions based on equivocal experimental results (Elashoff & Beal, 1976; Elashoff & Preston, 1977; Elashoff *et al.*, 1979). After each stage in a multistage experiment, one can either stop the experiment, if the classification of the compound as a carcinogen or noncarcinogen is clearly indicated by the data, or go on to the next stage, if the evidence regarding carcinogenicity is equivocal. By using a multistage design, it should be possible to lower the false-positive and false-negative rates by reducing the number of decisions made on the basis of equivocal experimental results.

Elashoff *et al.* (1979) considered a particular two-stage design in some detail; they compared the operating characteristics of a screening programme based on their two-stage design with those of a typical one-stage design, in which a control group of 50 animals is compared to two groups, each of 50 animals, exposed to different dose levels of the test compound. The first stage of their two-stage design is identical to the one-stage design, except that each group has only 35 animals. If the results of the first stage lead to a clear-cut classification of the test compound, then the experiment is terminated. If, however, the first stage gives equivocal results, then the second stage is conducted. The second stage compares a control group of 35 animals to a single exposed group of 35 animals. The evaluation of the second stage is simplified by the fact that any target organs have been identified in the first stage. Thus, the analysis of the second stage is restricted to the identified target organs, and nominal significance levels can be increased somewhat. Elashoff *et al.* (1979) showed that, using their two-stage design, a particular (hypothetical) large pool of chemicals could be screened in a shorter time and with a greater savings of animals than using the one-stage design. These savings were accomplished without increasing the false-positive or false-negative rates of the screening programme.

Although more compounds may be tested in a given period of time using such designs, individual multistage experiments will take longer to complete than single-stage experiments whenever a decision is not reached at the end of the first stage. As with the single-stage designs, moreover, it is still possible that a clear-cut classification concerning carcinogenicity may not be obtained following the completion of a two-stage test. For these reasons, multistage designs are more likely to find application in large-scale bioassay programmes involving many compounds than in studies in which a single substance is to be assessed.

3.4 Designs for dose-response studies

While a screening study is a useful tool for assessing carcinogenic potential on a qualitative basis, a dose-response study is required in order to describe the characteristics of the dose-response curve in more quantitative terms. This is particularly important when an identified carcinogenic hazard cannot be removed readily from the environment.

The evaluation of dose-response data is often done following a specified period of exposure, such as 24 months for rats and 18 months for mice. More generally, the probability of tumour induction may be considered to be a response surface $P(t, d)$ depending on the exposure time t and the dose d (Figure 3.3). Generally, the response probability $P(t, d)$ may be expected to increase with both dose d and time t.

The dose-response curve thus depends on the exposure time, as illustrated in Figure 3.4 by the data on liver and bladder tumour induction as a result of exposure to 2-acetylaminofluorine (2-AAF) from the ED_{01} study conducted by the US National Center for Toxicological Research (Littlefield *et al.*, 1980a). This study included large groups of mice exposed to increasing doses of 2-AAF for periods of 18, 24 and 33 months. While the general shapes of the dose-response curves for either bladder or liver tumours are similar at each of the three exposure times, the dose-response curve does vary somewhat with time. The results also indicate that the risk at all dose levels increases with the period of exposure to 2-AAF.

Fig. 3.3 The probability of tumour induction $P(t, d)$ represented as a response surface depending in time t and dose d (after Krewski *et al.*, 1983)

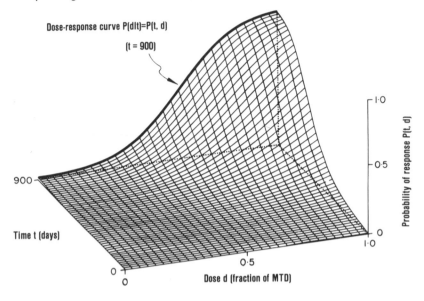

Fig. 3.4 Dose-response curves for bladder and liver tumours induced by 2-AAF following 18, 24 and 33 months' exposure (after Littlefield *et al.*, 1980a)

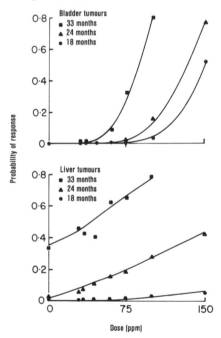

In the absence of any prior information on the shape of the dose-response curve, any reasonable series of increasing nontoxic dose levels may be used. Generally, the more dose levels, the greater the resolution of the dose-response curve. The largest dose-response study conducted to date, the ED_{01} study at the US National Center for Toxicological Research (Cairns, 1980), involved seven doses of 2-AAF ranging from 30–150 ppm in the diet, and an untreated control group, with between about 100 and 1700 Charles River BALB/c female mice at each dose. In total, 24 192 mice were used. Although the lowest and highest doses were separated by a factor of only five, the shape of the dose-response curves for bladder and liver tumours was well determined. (This compact design was possible due to the conduct of a smaller, pilot study.)

Another study of comparable size was commissioned by the UK Ministry of Agriculture, Fisheries and Food (Peto *et al.*, 1984). This involved a total of 5000 rodents (mice, hamsters and rats) administered different nitrosamines in the drinking-water. The main experiment involved one control and 15 dose groups of 48 males and 48 females each, with doses ranging from 0.033 to 16.896 ppm. In Section 4.3, we give the data from this study on the occurrence of pituitary tumours in male Colworth rats exposed to *N*-nitrosodimethylamine.

The two experiments described above are atypical because of their magnitude. Cost considerations normally dictate the conduct of smaller studies involving one control and three or four dose levels (see Table 3.1). Although less definitive, even these smaller studies can provide valuable information on the shape and nature of the dose-response relationship.

Because human exposure to most environmental carcinogens is low, data from dose-response studies are sometimes used to estimate the probability of tumour induction at low-dose levels. Since direct estimates of small probabilities are not feasible with small experiments (Kalbfleisch *et al.*, 1982), indirect estimates based on the extrapolation of results obtained at high doses are necessarily obtained using a particular dose-response model (Krewski & Van Ryzin, 1981; Kalbfleisch *et al.*, 1983). (Possible models are discussed in detail in Section 6.2.)

The problem of selecting the most suitable experimental design for puposes of low-dose extrapolation has been the subject of several investigations. However, the optimal design depends to some extent both on the assumed model and on the criterion used to compare competing designs. Experimental designs which minimize the asymptotic variance of maximum-likelihood estimates of risk, using a variety of three-parameter response models, have been investigated by Krewski *et al.* (1984b). With only three parameters in the dose-response model, the optimal design will involve only three dose levels (Chernoff, 1972). In Table 3.9 we give the response rates at three dose levels for different mathematical dose-response models (details of these are given in Section 6.2). This table shows the corresponding proportional allocation of animals which would lead, for the given three-parameter model, to the most precise estimation of the dose corresponding to a risk of 10^{-5}. Note that, for many agents, the response rate at the high dose in the optimal design may exceed that found at the MTD.

Taking the MTD into consideration, the optimal designs were generally found to involve three treatment groups, including one group at the MTD and one control

Table 3.9 Three-dose unrestricted optimal design for low-dose extrapolation to a risk of 10^{-5} under the three-parameter probit, logit and Weibull models

Model	Optimal response rates[a]			Optimal allocation (%)[b]		
	p_0	p_1	p_2	c_0	c_1	c_2
Probit	0.01	0.129	0.953	13	52	35
	0.05	0.222	0.963	19	45	36
	0.10	0.296	0.968	21	43	36
Logit	0.01	0.154	0.931	10	47	43
	0.05	0.244	0.945	15	40	45
	0.10	0.316	0.953	16	38	46
Weibull	0.01	0.199	0.976	11	57	32
	0.05	0.296	0.980	18	49	33
	0.10	0.370	0.983	20	46	34

[a] Response rates at the three optimally selected doses (In all cases, p_0 was found to correspond to the spontaneous response rate)
[b] Percentage of animals to be allocated to each of the three optimal doses

group. The dose given to the intermediate dose group depends on the curvature of the dose-response curve, with greater curvature requiring a larger fraction of the MTD. The allocation of animals among the dose groups depends on a number of parameters, including the acceptable risk and background response rate. However, a 1:2:1 allocation, with half of the animals on the low dose and the other half divided evenly between the control and high-dose groups, appears to result in a reasonably efficient design. An interesting property of these designs is that they are practically independent of the particular model assumed. In addition, designs for which dose placement and animal allocation differ moderately from those of the optimal design maintain high efficiency. One disadvantage of such optimal designs is that one has to know something of the shape of the dose-response curve in order to determine the optimal design.

For this reason, both Krewski *et al.* (1984b) and Portier and Hoel (1983b) have considered the efficiencies of suboptimal designs that may be expected to perform reasonably well in a variety of situations. In spite of the different approaches and models, both of these investigations have yielded similar conclusions. The former investigators proposed a design with one control and three equally spaced groups, at 0, $\frac{1}{3}$, $\frac{2}{3}$ and 1 times the MTD or at 0, $\frac{1}{4}$, $\frac{1}{2}$ and 1 times the MTD. Both 1:1:1:1 and 1:2:2:1 animal allocations for these doses were found to perform well. Portier (1981) has recommended a design with similar dose levels and a 2:3:3:2 animal allocation. These designs are again similar to those recommended by Gaylor *et al.* (1985a), who attempted to obtain the tightest possible confidence limits rather than to minimize the variance of the low-dose risk estimates.

The question of which mathematical model to use is of great concern in low-dose extrapolation (see Section 6.1). Both Chambers and Cox (1967) and Crump (1982) have developed optimal designs for discriminating between two specified dose-response models. (Related results for assessing the goodness-of-fit of multistage models have been given by Portier & Hoel, 1984a.) Unfortunately, their results are not particularly encouraging. Even with such optimal designs, several thousands of animals are required to permit reasonable discrimination between two plausible models.

Because of the uncertainty regarding the shape of the dose-response curve in the low-dose region, some form of linear extrapolation is often employed in an attempt to obtain an upper limit on risk. Low-dose linearity may occur with carcinogens that interact directly with genetic material (Hoel et al., 1983). These authors also demonstrate that nonlinearity at higher doses may be due to saturation effects in the metabolic activation process. Another argument leading to low-dose linearity is that of dose-wise additivity (Crump et al., 1976). Under this hypothesis, spontaneous lesions may be modelled as arising from an effective 'background dose' of the test agent.

A simple procedure for linear extrapolation (Van Ryzin, 1980) involves extrapolating linearly from the 1% or 10% response point, based on a suitable model. An important feature of the above designs is that they are also nearly optimal for this procedure, at least in terms of minimizing experimental error. Other designs may be required in order to minimize systematic errors in model specification as well as random experimental error (Lawless, 1984).

3.5 Designs for studies of mechanism

Designs for studies that examine certain hypotheses concerning the mechanisms of carcinogenesis generally need to be tailored to a particular experiment. Because most past data have been collected for the purpose of screening chemicals, however, the statistical procedures required to assess adequately the data from such experiments still remain to be established. Nonetheless, some general observations on the design of specific types of studies are given below.

Two-generation studies may be used in cases in which the test agent may exert its effect after exposure *in utero* (Grice et al., 1981). Although prenatal exposure to certain nitroso compounds can induce tumours in young animals (Ivankovic, 1973), other compounds may require two-generational exposure in order to exert their full carcinogenic potential. The latter form of exposure was long thought to be necessary to obtain bladder lesions with saccharin (Arnold et al., 1983a), although recent results (Shubik, 1985) have demonstrated that exposure from birth onwards may be sufficient. Generally, however, in order to distinguish between effects induced prenatally and postnatally, two-generation dosing regimens need to be included in the study protocol (Figure 3.5).

Another form of study of mechanisms addresses the multistage nature of carcinogenesis. In particular, there is now a substantial body of literature which suggests that tumour formation is a multistage process (Pitot & Sirica, 1980; Clayson, 1981). Initiation is thought to involve direct interaction between the proximate carcinogen and cellular DNA, although subsequent promotional events may be required for tumour development. In order to test a particular initiator/promoter pair in this model system, however, several dosing regimens involving various applications of the initiating and promoting agent may be required (Williams et al., 1981). The primary endpoint in most two-agent designs, such as those described in Figure 3.6, is the development of

Fig. 3.5   Possible dosing regimens in a two-generation cancer bioassay (after Bickis &
Krewski, 1985)

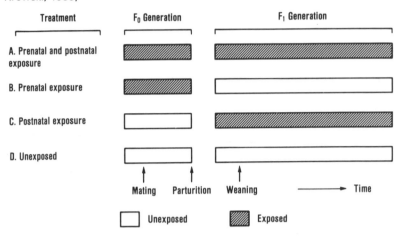

papillomas. More complicated dosing regimens, often using more than two test agents,
are required to investigate theories regarding different stages of promotion in
papilloma formation and the number of genetic alterations required to transform a
normal cell to a carcinoma (Hennings, 1986).

A third type of study design would be necessary to investigate the effect of
discontinued exposure on the development of neoplastic or preneoplastic lesions (Day

Fig. 3.6   Possible dosing regimens in an initiation/promotion study (after Bickis & Krewski,
1985)

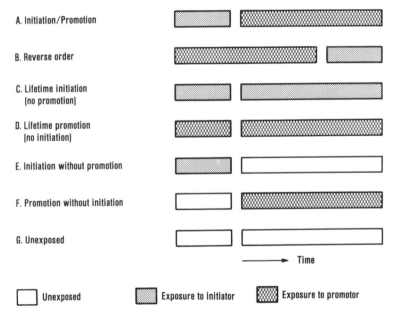

Fig. 3.7   Possible dosing regimens in a discontinued exposure study (after Bickis & Krewski, 1985)

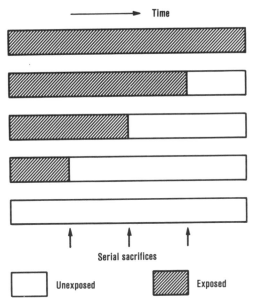

& Brown, 1980; Arnold *et al.*, 1983b). In this case, dosing regimens involving exposure to the test agent followed by a return to control conditions would be employed, as illustrated in Figure 3.7. Both exposure and non-exposure periods of varying duration could be employed to explore the reversibility hypothesis in greater detail, as in the ED_{01} study (Cairns, 1980). The results of that study indicated that bladder neoplasms induced by 2-AAF occurred early in the study but were dependent on continuous exposure. In addition, moderate or severe hyperplasia apparently regressed when 2-AAF feeding was discontinued (Littlefield *et al.*, 1980b). In contrast, liver neoplasia occurred late in the study and did not require continuous exposure to 2-AAF (Littlefield *et al.*, 1980).

While the preceding designs focus on different aspects of the effects of exposure to a single toxicant, it is well known that the effect of certain carcinogens is much greater in combination than singly, as is the case with lung cancer mortality observed among asbestos workers who also smoke (Hammond *et al.*, 1979). Because of the ever-increasing number of potentially toxic agents present in the human environment, mixed exposures need to be evaluated for safety (Freundt, 1982). In order to explore the possibility of interactions between chemicals and other test agents, experiments involving exposures to mixtures of toxicants are required in order to evaluate synergistic or antagonistic potential (Wahrendorf & Brown, 1980; Wahrendorf *et al.*, 1981; Abdelbasit & Plackett, 1982; Métivier *et al.*, 1984).

In designing such multifactorial studies, investigators should clearly identify and distinguish the relevant factors which can be or must be manipulated in the design. Such factors include those which are essential for biological and technical reasons, such as sex, batch of animals, solvents of the exposure, or others. The different factors that are intended to be manipulated deliberately must then be defined very concisely. These factors may be two chemicals, one chemical and one radiation exposure, or individual constituents of a commonly-found, complex mixture, such as betel quid or tobacco-smoke condensate.

The number of levels at which each factor can be investigated meaningfully has to be decided upon. A complete factorial design would then include as many experimental groups as the product of the number of levels of all the considered factors. Note that an untreated or appropriately vehicle-treated control level of each factor has to be included. For example, Métivier *et al.* (1984) report an experiment in which four dose levels (that is, unexposed and three doses) of exposure to an aerosol of [^{239}Pu]-plutonium oxide were combined with two (that is, vehicle treatment and one dose) levels of intratracheal benzo[a]pyrene instillation, resulting in $4 \times 2 = 8$ experimental groups.

In general, however, the number of experimental groups and the corresponding number of animals required in such studies may be large, even if only two factors are being studied. In order to permit a clear analysis of the individual and combined effects, an unexposed level should be included for each factor, and all possible combinations should be maintained in the experimental design.

3.6 Histopathological analysis

Histopathological analysis forms perhaps the most critical component of the long-term carcinogenicity bioassay. Since important conclusions concerning the determination of carcinogenicity are based on comparisons of patterns of tumour occurrence between exposed and unexposed animals, it is imperative that pathological examinations follow the most stringent standards of quality, uniformity and objectivity (Ward *et al.*, 1978; Ward & Reznick, 1983). This requires adherence to proper procedures at the time of necropsy including gross tissue examination, the selection of tissues to be examined, the sectioning and histological preparation of tissue samples and the evaluation of tumour morphology.

Although many lesions may be detected readily following gross examinations, it is also essential that all tissues of concern be subjected to histological examination, since many microscopic lesions not visible at the time of necropsy can be detected using modern histological procedures (Kulwich *et al.*, 1980). For example, more than 50% of the neoplastic lesions present in organs such as the thymus, lung, adrenal, Harderian gland and urinary bladder were not apparent during the gross examination in studies done at the US National Center for Toxicological Research (Frith *et al.*, 1980).

Generally, past practice in evaluating histopathological data has been to indicate only the presence or absence of particular lesions at specified sites. This can be refined in two ways. First, since many carcinogenic effects progress through a series of stages involving minimal to advanced neoplastic changes, it is often possible to assign a grade to such changes. For example, tumour status might be categorized on a scale from 0 to 5, indicating the absence of a lesion, or minimal to severe effects. Second, morphometric techniques may be used to quantify the extent of such changes (Kuiper-Goodman *et al.*, 1976). The objective of both grading and morphometric analysis is to provide more detailed information on which to base statistical analyses, and, hence, to obtain stronger conclusions concerning the carcinogenic potential of the test agent. Nonetheless, these two procedures require considerable additional time and effort, and are thus not yet widely applied.

The question whether pathological evaluation of histologically prepared tissues should be performed without knowing the treatment group of the subject has been the topic of considerable debate over the years (Weinberger, 1973, 1979; Fears & Schneiderman, 1974). From the statistical point of view, it is desirable that pathology, both gross and histological, should be done 'blind'. This does not mean that the pathologist should be given simply a numbered slide and asked to identify the lesions present. All tissues of the animal should be evaluated as a unit, and the entire clinical history of the animal should be available to the pathologist. Nowhere on this record, however, should there be any indication of the animal's treatment group; in fact, the pathologist need not know if the individual animals he is diagnosing come from different treatment groups. All animals should be considered equivalent, except from what he can observe. For this reason, the animals should be presented in a random order.

The argument is sometimes raised that, unless the pathologist knows which is the high-dose group, he has difficulty diagnosing effects since he does not know what kind of lesion to expect from the treatment. In this case, a pilot experiment might be advisable, or else a satellite group of high-dose animals may be included for the pathologist to examine. In order to obtain an unbiased assessment, however, the results from these animals would not be used in assessing the significance of those from the main study.

Another possible approach is to have the pathologist examine a selection of control and high-dose lesions in order to familiarize himself with the nature of the lesions to be diagnosed. After this initial examination, these same slides would then be re-read 'blind' and in a random order along with those from the remaining animals.

While blind reading is clearly preferable, the avoidance of 'diagnostic drift' or time-related changes in the evaluation of the slides is perhaps of more serious concern. Thus, the difference between blind studies and those read non-blind but in a random order may be less important than the difference between studies read non-blind but in random order and those read non-blind in some systematic order.

Because of the large number of tumours examined in a typical bioassay, methods for reducing the histopathological workload by examining only a sample of the slides have been considered by Fears and Douglas (1977a,b). According to their proposal, a complete set of slides would be read only if it appeared necessary on the basis of the sample results. Although this offers some potential savings in costs, current practice is to subject all slides to a thorough histopathological examination in order to obtain as much information as possible.

3.7 Recording of experimental data

An adequate system for collecting, processing, reporting and storing large amounts of data is an essential part of the design of any long-term bioassay and is most easily organized using a computerized data storage and retrieval system (Naylor, 1978; Cranmer et al., 1978; Konvicka et al., 1978; Felsky et al., 1979; Lawrence et al., 1979; Herrick *et al.*, 1983). This could involve separate subsystems for the different data sets generated during the conduct of a long-term study, such as those for feed and water consumption, body weight, haematology and pathological findings. Another approach is to use one master system to acquire and integrate data from these different sources and to integrate them into a common data base. This latter approach is advantageous in laboratories where the studies are all of a similar nature, but it is less flexible in situations where the experiments are more variable.

The pathologist should record data in a systematic way so that the information is readily transferable to the computer *via* a system suitable for this purpose (Frith *et al.*, 1977; Naylor, 1978; Faccini & Naylor, 1979). Care should be taken to ensure that the same lesion is always described in the same way, and that quantitative assessments are given, where possible, of size, number and severity of observed lesions. Organs for which sections are not available, or which are too autolysed for examination, should be clearly marked. Unless indicated to the contrary, failure to mention a particular lesion described in another animal should always imply that the lesion was not present. Methods by which the pathologist's report and the statistical analysis are both generated by computer from the same input data source are to be preferred over systems in which the report is dictated and the extraction of data represents the first step in statistical analysis.

Systems are now available by which the pathologist can enter data directly *via* a computer terminal. However, it is not clear that this is the most practical method, as it requires the pathologist to move continually from the microscope to the terminal. Roe and Lee (1984) have developed a complete system for recording, reporting and statistical analysis of histopathological data from animal studies. Pathologists may enter data directly into the computer, or onto *pro forma* (which can be modified according to requirements) which are then subsequently processed for computer input.

Care should be taken also to guard against variation in standards of diagnosis both among different pathologists and in different time periods for each pathologist. Where the experiment is so large that more than one pathologist is required, it is important to avoid systematic differences in recording lesions which may result in a misleading indication of treatment-related effects. One simple procedure is to ensure that each pathologist has the same proportion of animals from each treatment group. A better method may be to assign different pathologists to different tissues across the board. Whatever approach is adopted, the identity of the pathologist responsible for a specific diagnosis should be recorded.

It is also important to guard against changes in standards with time. For example, if a pathologist starts with the first animal in the first group and works systematically through to the last animal in the last group, dose-related trends in the severity of some lesions might not reflect any true treatment effect. To avoid bias due to diagnostic drift, the slides should be read in random order, or in an order which avoids any systematic tendency for all the animals in one dose group to be read in advance of the remaining animals. This is particularly important with respect to lesions which pose diagnostic difficulties or those scored according to their degree of severity. Whatever scheme is used, the date and time of the pathologist's determination should be recorded.

Care should also be taken to ensure consistency between the microscopic and macroscopic findings. Sections of suspect tumours noted macroscopically *post mortem* should be available for examination microscopically, and missing organs should be noted.

The data base for statistical analysis should include not only the pathologist's findings, but also the time at which the tumour was first noticed. This is feasible for visible or palpable lesions, and, in other cases, will correspond to the time of necropsy. For some experiments, the time at which lesions first reached different, specified sizes will also be recorded.

Ideally, it would be desirable also to record whether each tumour was the underlying cause of death. Sometimes this is an easy process, but more often it is not possible to give a reliable answer. One practicable compromise is to devise a four-point scale in which tumours are categorized as being 'definitely not', 'probably not', 'probably', or 'definitely' responsible for the death of the animal. A first estimate should be obtained at gross necropsy, with the possibility of revision later by histological examinations.

Finally, as we have noted, it is important that any relevant design features of the experiment are available for statistical analysis if required. These include not only details of the dose and the duration of treatments applied, but also details of cage placements, the name of the pathologist, and number of the animal and the time it was examined.

3.8 Summary and recommendations

In this chapter, we have provided an overview of the design of long-term animal experiments that examine potential carcinogenic effects of a test agent. It is clear from this discussion that the design of animal carcinogenicity experiments is a complex issue and that no hard and fast rules for their conduct can be laid out in advance. Nonetheless, a number of general statistical and other principles may be identified which should assist in the design of individual studies.

The preferred design in specific cases will be strongly dependent on the study's objectives. In this chapter, we have identified three broad categories of designs appropriate for screening, dose-response and mechanistic studies. Many of the studies conducted to date were intended to identify the presence or absence of carcinogenic activity in qualitative terms only; these fall into the first category. Screening studies of this type may involve only one or two dose groups and an unexposed control group. In order to maximize the chances of observing a carcinogenic effect in a relatively small population of test animals, moderate to high doses are used generally, with exposure continuing throughout the major portion of the animals' normal lifespan. The highest dose is however subject to the criteria for the MTD, particularly with respect to the requirements for minimal effects on body weight and overall survival (with the exception of increased mortality attributable to tumour induction). Lower doses are included often for confirmatory purposes and to ensure against the possibility that the high dose may have been exceeded. Commonly used sample sizes of about 50 animals per group are reasonably sensitive to effects involving 20% or more of the exposed population in cases where the background rate of occurrence of the lesion of interest is low.

A number of modifications to these conventional designs for screening studies, including the use of two-generation studies incorporating perinatal exposure, have been discussed in the literature. Multistrain studies involving the use of a variety of tester strains (generally with fewer animals per strain) have been proposed to increase the chances of selecting susceptible test groups. Sequential designs have been considered also as a means of decreasing the average cost and time involved in assessing a large number of test agents. These involve the conduct of a smaller study at the first stage, followed, when necessary for purposes of clarification in equivocal cases, by a second, confirmatory stage. While both of these latter proposals have merit, little past experience exists on which to base a sound evaluation of their properties. Thus, their use is recommended with caution.

Dose-response studies differ from screening studies in that additional dose levels are usually employed in an attempt to define more clearly the shape and nature of the dose-response relationship for the test agent. While elaborate studies of this type have been conducted with a limited number of compounds, such as 2-AAF and various nitrosamines, the use of one control and only three or four exposed groups is more common. Although optimal designs for low-dose risk assessment can be developed under specific parametric assumptions, they are fairly robust in terms of the efficiency of the resulting estimates, so that any reasonable design should be suitable.

Even more elaborate designs are required for studies of mechanism, depending on whether the objective is to study the effects of prenatal and postnatal exposure, to explore initiation/promotion hypotheses, to assess the impact of cessation of exposure on the carcinogenic process, or to evaluate the effects of joint exposures to different agents. All these studies require a variety of specialized treatment combinations in order to isolate the effects of interest; in addition, these assays need the use of one or more specialized control groups for comparison purposes.

Regardless of the nature of the study, there are a number of fundamental statistical principles that must be taken into account in developing a suitable experimental design. Randomization is essential in order to ensure unbiased comparisons between the treatment groups of interest and to provide a basis for valid statistical inference in terms of probability statements concerning evidence against the null hypothesis of no effect. Replication is essential in order to provide an estimate of experimental error against which to gauge the significance of any apparent related effects. Stratification may be used to decrease the magnitude of the experimental error by making treatment comparisons within homogeneous subgroups in order to increase the precision of the overall analysis.

The concept of an experimental unit is essential to the understanding of experimental design. While, in the past, the individual animal has been treated as the basic unit of information for purposes of statistical analysis, clustering due to cage or litter effects or environmental gradients existing within the laboratory suggests the possibility of one unit being comprised of two or more animals, falling within the same cluster. Since the available evidence on this effect is somewhat equivocal, it requires clarification.

More attention needs to be paid to the development of randomization schemes for the location of cages in carcinogenicity testing. While complete randomization has a desirable element of simplicity and offers protection against positional effects, it also poses potential problems in terms of cross-contamination of the treatment groups. Cross-contamination can be avoided by assigning animals on the same treatment to the same rack or column, although consideration should then be given to adjusting for possible environmental effects at the analysis stage. The possibility of rotating cage positions may be helpful in this regard but the most suitable rotation scheme is unclear.

Care needs to be taken at the design stage to develop a good system for the recording of experimental data. Different computer systems may be used for this purpose and have the advantage of increased accuracy and speed over manual systems. Special consideration also needs to be given to the standardization of the diagnostic criteria used in reviewing histopathological data and the development of an unbiased system for this review. In addition to the pathology data itself, it is recommended that, whenever feasible, some attempt be made to determine cause of death.

4. SOME ILLUSTRATIVE DATA SETS

CHAPTER 4

SOME ILLUSTRATIVE DATA SETS

In this chapter we give some detailed data sets which are used as examples in the subsequent chapters.

4.1 Bioassay of 1,2-dichloroethane

The first data set is taken from an NCI (USA) bioassay of 1,2-dichloroethane (National Cancer Institute, 1978). For our purposes, we have considered only data on the occurrence of alveolar/bronchiolar adenomas of the lung in female B6C3F1 mice in three groups – a high-dose group, a low-dose group and a control group, with originally 50, 50 and 40 animals, respectively. Dosage was continued for 78 weeks and terminal sacrifice carried out after 90 weeks. The lungs of all animals, apart from two autolysed animals in the high-dose group, were examined histopathologically; the data are summarized in Table 4.1. Although, in this experiment, no attempt was made to record the context of observation of the tumours, there appears to be agreement that these lung adenomas can be considered as incidental findings, that is, they did not contribute to the deaths of the animals.

4.2 Skin-painting experiment with cigarette- and cigar-smoke condensates

The second data set is taken from a mouse skin-painting study carried out by the Tobacco Research Council, UK (Lee, 1979). This study included six groups of 100 Carworth Swiss female mice, three painted with condensate from standard cigarettes at dose levels of 180, 136 and 103 mg per week and three painted with condensate from standard cigars at dose levels of 103, 78 and 59 mg per week. The dose levels were chosen from past experience so as to give a similar range of responses on a rising part of the dose-response curve, to be at equal steps on a logarithmic scale and to allow some comparison between identical doses of the two carcinogens. Condensates were applied continuously three times per week and continued for 100 weeks, at which time all surviving mice were killed. The data are presented in Table 4.2, giving time to visible skin tumour (if applicable) or time to death or terminal kill for all animals.

4.3 Pituitary tumours in rats treated with various doses of *N*-nitrosodimethylamine (NDMA)

A large animal experiment with a total of 5000 rodents was conducted by the British Industrial Biological Research Association (Peto *et al.*, 1984). The study was designed

Table 4.1   Alveolar/bronchiolar adenomas of the lung in female B6C3F1 mice treated with 1,2-dichloroethane by gavage

Control group			Low-dose group			High-dose group		
Time (weeks)	No. of deaths	No. of animals with tumours	Time (weeks)	No. of deaths	No. of animals with tumours	Time (weeks)	No. of deaths	No. of animals with tumours
22	1		46	1		11	3	
56	1		61	1		12	4	
60	1		66	1		14	1	
65	1		70	1		31	1	
77	1		73	1		36	1	
79	1		78	1		44	1	
85	1		81	1		57	1	
90	33	2	82	2		62	2	1
			86	2		63	2	2
			87	1		65	4	2
			88	3		66	1	
			90	35	7	67	2	1
						68	4	
						69	2	1
						70	3	1
						71	1	
						73	4	2
						74	2	1
						75	2	1
						76	1	1
						77	3	
						79	1	
						80	1	1
						90	1	1
Total	40	2		50	7		48	15

to investigate the dose-response relationship for the carcinogenicity of different nitrosamines administered in drinking-water.

In the main part of the study, for each of the two nitrosamines investigated, N-nitrosodimethylamine (NDMA) and N-nitrosodiethylamine (NDEA), 16 treatment groups were considered: an untreated control group was given 0 ppm (group 1), and the 15 treated groups were given 0.033 ppm (group 2), 0.066 ppm (group 3), 0.132 ppm (group 4), 0.264 ppm (group 5), 0.528 ppm (group 6), 1.056 ppm (group 7), 1.584 ppm (group 8), 2.112 ppm (group 9), 2.640 ppm (group 10), 3.168 ppm (group 11), 4.224 ppm (group 12), 5.280 ppm (group 13), 6.336 ppm (group 14), 8.448 ppm (group 15) and 16.896 ppm (group 16). Inbred Colworth rats were used and, for each compound and each sex, the control group comprised 192 animals and each treated group, 48.

The context of observation was recorded for all tumours found. Subsequently, we give the data for the occurrence of pituitary tumours in male rats given NDMA. Table 4.3 lists, for all 16 groups, the time to death for each animal and the context of observation for pituitary tumours. Time to death is recorded in days, whereas the

Table 4.2 Time (in weeks) to skin tumour or death (terminal kill) for six groups of Carworth Swiss female mice

Group A: Cigarette condensate 180 mg/week
(a) 43 animals with skin tumour at:

24	24	29	31	32	34	34	38	41	41
41	41	43	48	48	49	51	57	58	59
61	61	62	62	62	63	64	65	66	66
68	72	72	75	76	79	79	81	84	85
88	88	100							

(b) 57 animals without skin tumour died at:

4	8	8	8	8	8	12	12	12	12
17	17	17	20	20	20	22	28	28	28
33	36	43	47	50	50	51	51	51	52
54	54	56	58	62	62	64	64	65	66
67	68	69	71	71	71	71	72	75	78
81	81	82	86	87	96	96			

Group B: Cigarette condensate 136 mg/week
(a) 36 animals with skin tumour at:

32	38	46	46	50	61	61	63	64	66
67	69	71	71	72	72	75	75	75	79
79	79	80	81	81	81	82	84	84	85
88	88	89	90	93	98				

(b) 64 animals without skin tumour died at:

5	6	8	12	12	12	12	12	16	16
16	17	19	19	20	22	23	28	31	36
36	38	40	40	41	44	47	50	52	53
53	54	59	64	65	66	68	69	70	71
74	74	76	80	84	84	84	86	88	89
91	93	95	95	96	98	98	98	100	100
100	101	101	101						

Group C: Cigarette condensate 103 mg/week
(a) 13 animals with skin tumour at:

20	27	35	47	50	64	65	84	84	88
93	98	100							

(b) 87 animals without skin tumour died at:

7	8	10	11	12	14	16	16	16	16
16	17	17	20	20	23	23	24	25	29
31	34	34	43	45	46	47	48	49	50
55	57	59	60	64	65	66	68	68	72
72	74	74	74	74	75	77	78	79	79
79	80	80	81	82	83	83	84	84	84
84	84	85	88	89	91	91	91	93	94
95	95	95	96	96	97	100	100	101	101
101	101	101	101	101	101	101			

Group D: Cigar condensate 103 mg/week
(a) 37 animals with skin tumour at:

38	43	43	48	49	50	50	50	50	57
57	58	58	58	61	62	64	64	66	66
71	71	72	72	72	75	76	79	79	81
84	84	86	88	93	93	99			

Table 4.2 *Contd.*

(b) 63 animals without skin tumour died at:

5	6	7	7	8	8	8	8	8	8
8	8	11	11	11	12	16	20	20	20
22	25	25	28	28	28	30	34	35	36
36	38	40	54	55	55	56	57	59	61
61	61	63	63	64	65	68	68	69	73
75	79	81	82	82	84	90	91	92	92
97	100	101							

Group E: Cigar condensate 78 mg/week

(a) 31 animals with skin tumour at:

38	41	41	49	54	57	57	58	61	64
64	64	71	71	72	72	75	76	76	76
79	79	82	84	85	86	93	93	93	98
100									

(b) 69 animals without skin tumour died at:

7	7	7	8	8	8	8	8	8	12
13	14	16	16	17	19	20	20	20	25
25	27	29	33	33	37	39	40	41	43
44	44	47	50	50	50	53	56	57	63
63	66	70	71	75	79	79	81	81	81
82	85	87	88	89	94	96	97	98	98
98	100	100	100	100	101	101	101	101	

Group F: Cigar condensate 59 mg/week

(a) 21 animals with skin tumour at:

24	31	38	39	46	46	53	54	58	66
71	73	75	79	79	80	81	85	93	94
96									

(b) 79 animals without skin tumour died at:

7	12	12	12	12	12	14	14	16	16
16	16	16	17	19	20	20	25	25	28
33	36	47	48	49	50	50	53	57	57
59	60	62	63	68	69	69	70	78	84
84	84	85	85	86	89	89	89	90	90
90	91	91	92	92	94	97	97	98	98
99	100	100	101	101	101	101	101	101	101
101	101	101	101	101	101	101	101	101	

context of observation is given as:

  0 no pituitary tumour found,
  1 incidental,
  2 probably incidental,
  3 probably fatal,
  4 fatal,
−3 totally cannibalized or autolysed; cause of death not ascertainable,
−2 head cannibalized or autolysed; presence or absence of pituitary tumour not ascertainable but death known *not* to be caused by pituitary tumour.

Table 4.3 Group number, time to death (in days) and context of observation of pituitary tumours in male Colworth rats exposed to *N*-nitrosodimethylamine (NDMA)

1	4	−3	1	793	0	1	931	0	1 1029	0
1	50	−3	1	795	0	1	932	4	1 1036	0
1	197	0	1	796	0	1	933	4	1 1037	2
1	260	0	1	796	0	1	934	0	1 1039	0
1	297	0	1	797	1	1	937	0	1 1044	4
1	302	0	1	799	1	1	937	0	1 1050	4
1	373	0	1	806	0	1	944	0	1 1050	4
1	415	0	1	812	0	1	944	0	1 1051	1
1	471	0	1	823	1	1	952	0	1 1055	0
1	476	−2	1	826	−2	1	954	0	1 1058	0
1	496	0	1	835	0	1	955	0	1 1062	0
1	502	0	1	837	0	1	959	0	1 1063	4
1	523	0	1	840	0	1	964	0	1 1066	0
1	534	4	1	840	0	1	965	1	1 1070	0
1	604	0	1	846	4	1	965	1	1 1073	1
1	607	−3	1	847	0	1	966	1	1 1075	0
1	617	0	1	848	0	1	966	0	1 1078	1
1	618	0	1	851	0	1	968	0	1 1078	0
1	623	0	1	856	4	1	973	0	1 1079	4
1	637	0	1	858	4	1	973	0	1 1079	0
1	652	0	1	864	0	1	973	1	1 1087	1
1	652	−2	1	867	0	1	974	0	1 1092	4
1	672	0	1	871	0	1	975	1	1 1097	0
1	676	0	1	872	0	1	977	0	1 1100	0
1	679	3	1	875	0	1	981	0	1 1101	1
1	688	0	1	881	0	1	981	4	1 1102	0
1	700	0	1	884	4	1	986	0	1 1103	0
1	700	0	1	884	0	1	988	0	1 1113	0
1	713	0	1	885	0	1	989	0	1 1117	0
1	716	0	1	886	4	1	991	0	1 1125	0
1	721	4	1	889	0	1	994	0	1 1135	1
1	723	0	1	895	1	1	995	0	1 1135	1
1	724	4	1	897	0	1	996	0	1 1147	1
1	726	0	1	897	0	1	999	4	1 1149	0
1	727	−3	1	902	0	1	999	0	1 1151	0
1	732	0	1	904	0	1	999	0	1 1156	4
1	734	0	1	904	4	1 1003	0	1 1159	0
1	734	0	1	905	0	1 1003	0	1 1165	0
1	747	0	1	908	1	1 1003	0	1 1182	1
1	750	0	1	910	0	1 1006	0	1 1184	0
1	756	0	1	916	1	1 1009	0	1 1187	4
1	764	4	1	918	0	1 1013	0	1 1213	0
1	765	4	1	919	0	1 1015	0		
1	766	0	1	919	0	1 1022	0		
1	773	0	1	919	1	1 1024	0		
1	776	0	1	921	1	1 1024	0		
1	777	4	1	921	4	1 1024	0		
1	778	0	1	922	0	1 1026	0		
1	782	0	1	923	0	1 1027	0		
1	785	0	1	925	1	1 1027	0		

Table 4.3 *Contd*.

2	380	0	3	529	0	4	343	−3	5	317	0
2	447	0	3	646	0	4	458	−2	5	619	0
2	454	0	3	646	0	4	493	0	5	621	0
2	512	0	3	677	0	4	520	0	5	626	0
2	552	4	3	679	0	4	559	0	5	632	0
2	623	0	3	693	0	4	623	0	5	677	−2
2	696	0	3	714	0	4	664	0	5	688	4
2	703	4	3	724	0	4	680	0	5	706	0
2	748	0	3	728	0	4	709	0	5	709	0
2	753	0	3	751	4	4	714	0	5	733	0
2	785	0	3	764	0	4	761	0	5	736	1
2	835	0	3	766	0	4	767	4	5	742	0
2	837	0	3	783	4	4	779	0	5	749	0
2	844	0	3	784	0	4	808	1	5	776	0
2	863	4	3	819	4	4	828	0	5	791	4
2	873	0	3	824	0	4	833	4	5	798	0
2	901	0	3	828	0	4	848	1	5	804	0
2	907	4	3	830	0	4	852	0	5	809	0
2	913	4	3	835	0	4	879	4	5	826	0
2	916	4	3	854	0	4	908	0	5	828	0
2	917	4	3	874	0	4	928	0	5	828	0
2	926	0	3	874	0	4	939	1	5	832	4
2	967	0	3	882	4	4	945	1	5	842	0
2	974	0	3	917	0	4	952	0	5	843	0
2	982	0	3	924	1	4	966	0	5	895	0
2	984	0	3	935	0	4	969	0	5	900	0
2	985	0	3	960	0	4	973	0	5	903	0
2	996	0	3	960	0	4	980	4	5	915	0
2	997	0	3	962	0	4	986	1	5	923	4
2	997	0	3	987	1	4	999	0	5	960	4
2	1002	1	3	988	0	4	1003	1	5	967	0
2	1010	3	3	989	4	4	1003	0	5	976	0
2	1016	0	3	995	1	4	1004	0	5	986	0
2	1029	1	3	996	4	4	1018	0	5	988	4
2	1033	1	3	1013	4	4	1020	0	5	1007	0
2	1036	0	3	1021	0	4	1021	1	5	1015	1
2	1048	1	3	1044	1	4	1023	0	5	1046	0
2	1065	0	3	1045	0	4	1029	0	5	1049	0
2	1081	0	3	1048	1	4	1056	4	5	1065	1
2	1085	3	3	1059	0	4	1068	0	5	1071	0
2	1095	1	3	1066	4	4	1071	0	5	1084	0
2	1102	0	3	1077	0	4	1084	0	5	1086	4
2	1110	0	3	1088	0	4	1084	0	5	1092	0
2	1118	1	3	1126	0	4	1090	0	5	1101	0
2	1130	3	3	1132	0	4	1095	1	5	1110	4
2	1135	0	3	1162	0	4	1114	1	5	1162	0
2	1148	0	3	1166	1	4	1121	0	5	1167	0
2	1234	0	3	1178	0	4	1188	0	5	1234	0

Continued

Table 4.3   *Contd*.

6	460	0	7	410	0	8	296	0	9	554	0
6	548	0	7	566	0	8	586	0	9	561	0
6	577	0	7	568	0	8	623	4	9	576	0
6	679	0	7	575	0	8	650	0	9	596	1
6	707	0	7	579	0	8	656	0	9	646	0
6	708	0	7	638	4	8	702	−2	9	671	0
6	713	0	7	656	0	8	728	0	9	683	0
6	750	0	7	678	0	8	736	0	9	684	0
6	777	4	7	680	0	8	737	4	9	709	4
6	786	4	7	708	0	8	740	0	9	730	0
6	821	0	7	712	0	8	767	0	9	742	4
6	825	0	7	714	4	8	771	0	9	743	0
6	838	0	7	719	0	8	771	4	9	744	0
6	848	0	7	723	0	8	776	0	9	766	0
6	874	4	7	742	0	8	779	0	9	773	0
6	898	4	7	777	0	8	800	0	9	821	0
6	898	4	7	797	4	8	802	0	9	823	0
6	901	1	7	805	0	8	811	4	9	838	0
6	916	4	7	806	4	8	816	4	9	861	0
6	929	0	7	830	0	8	838	0	9	864	0
6	948	0	7	842	0	8	843	3	9	870	0
6	951	4	7	844	0	8	851	4	9	880	4
6	952	4	7	846	0	8	858	1	9	891	0
6	953	4	7	858	4	8	872	4	9	921	0
6	961	0	7	863	4	8	889	0	9	925	1
6	961	0	7	888	0	8	892	1	9	930	0
6	966	0	7	909	4	8	895	0	9	936	4
6	973	0	7	917	0	8	897	0	9	940	0
6	994	0	7	937	0	8	901	0	9	947	0
6	1009	0	7	953	0	8	903	0	9	953	0
6	1010	0	7	954	0	8	908	0	9	955	0
6	1034	0	7	955	0	8	912	1	9	961	4
6	1041	1	7	956	0	8	932	0	9	970	0
6	1046	0	7	980	0	8	933	0	9	992	0
6	1053	0	7	982	1	8	946	0	9	996	1
6	1056	0	7	996	1	8	950	0	9	1009	1
6	1059	0	7	998	0	8	956	4	9	1028	0
6	1064	0	7	1000	4	8	994	0	9	1030	0
6	1073	1	7	1001	0	8	1003	4	9	1035	1
6	1073	1	7	1047	0	8	1020	1	9	1041	1
6	1085	1	7	1052	0	8	1028	0	9	1047	1
6	1092	4	7	1055	0	8	1037	0	9	1056	0
6	1094	1	7	1081	4	8	1045	0	9	1058	0
6	1120	0	7	1115	0	8	1063	0	9	1073	0
6	1126	1	7	1122	0	8	1071	4	9	1082	0
6	1127	1	7	1174	0	8	1078	0	9	1101	0
6	1137	0	7	1183	0	8	1082	1	9	1117	0
6	1218	0	7	1204	1	8	1089	1	9	1119	0

Table 4.3   *Contd*.

10	254	0	11	331	0	12	87	0	13	345	0
10	264	0	11	441	0	12	265	0	13	371	0
10	411	0	11	467	0	12	267	0	13	387	0
10	495	0	11	485	0	12	289	0	13	407	0
10	495	−2	11	526	0	12	393	0	13	412	0
10	516	0	11	527	0	12	419	0	13	435	0
10	554	0	11	533	0	12	470	0	13	442	0
10	569	0	11	559	0	12	499	0	13	470	0
10	601	0	11	596	4	12	506	0	13	476	0
10	627	0	11	600	0	12	516	0	13	477	0
10	633	0	11	614	0	12	539	0	13	483	0
10	660	4	11	645	4	12	555	0	13	483	0
10	660	4	11	651	0	12	581	0	13	485	0
10	663	−2	11	676	0	12	610	0	13	488	0
10	681	1	11	690	0	12	621	0	13	518	0
10	686	0	11	695	0	12	632	0	13	519	0
10	698	4	11	707	0	12	633	0	13	522	0
10	708	0	11	709	0	12	646	0	13	525	0
10	716	0	11	716	0	12	661	0	13	525	0
10	716	1	11	731	0	12	665	0	13	534	0
10	720	0	11	732	0	12	667	4	13	534	0
10	724	0	11	764	0	12	676	0	13	539	0
10	729	4	11	768	0	12	690	0	13	540	0
10	751	0	11	771	0	12	696	0	13	546	0
10	759	0	11	775	0	12	707	0	13	549	0
10	770	1	11	783	0	12	710	0	13	556	0
10	777	0	11	796	0	12	715	0	13	564	0
10	778	0	11	802	0	12	720	0	13	581	0
10	786	0	11	814	0	12	720	1	13	581	0
10	788	0	11	833	0	12	729	0	13	586	0
10	795	0	11	838	0	12	729	0	13	589	0
10	812	4	11	840	0	12	736	0	13	591	0
10	864	0	11	855	0	12	736	0	13	597	0
10	880	0	11	857	0	12	741	0	13	598	0
10	884	0	11	864	0	12	779	0	13	604	0
10	884	4	11	867	0	12	795	0	13	604	0
10	903	0	11	868	1	12	804	0	13	618	1
10	911	0	11	868	4	12	808	0	13	628	0
10	918	0	11	879	0	12	814	0	13	651	0
10	931	1	11	883	0	12	863	0	13	656	0
10	938	1	11	894	0	12	872	0	13	663	0
10	944	0	11	894	4	12	880	0	13	665	0
10	946	0	11	898	0	12	893	0	13	677	0
10	966	0	11	914	0	12	894	0	13	684	0
10	986	0	11	919	0	12	897	0	13	687	0
10	1015	0	11	919	4	12	904	0	13	700	0
10	1043	0	11	947	0	12	933	0	13	706	1
10	1086	0	11	957	0	12	1019	0	13	778	0

Continued

Table 4.3 *Contd.*

14	248	0	15	177	0	16	92	0
14	284	0	15	297	0	16	140	0
14	285	0	15	303	0	16	160	−2
14	372	0	15	309	0	16	164	0
14	376	0	15	316	0	16	168	0
14	421	0	15	362	0	16	169	0
14	428	0	15	364	0	16	170	0
14	432	0	15	365	0	16	180	0
14	441	0	15	371	−2	16	188	−2
14	471	0	15	383	0	16	189	0
14	474	0	15	388	0	16	194	0
14	477	0	15	390	0	16	200	0
14	482	0	15	399	0	16	200	0
14	498	0	15	399	0	16	205	0
14	498	0	15	405	0	16	206	0
14	498	0	15	416	0	16	208	0
14	499	0	15	417	0	16	210	0
14	505	0	15	423	0	16	210	0
14	507	0	15	433	0	16	211	0
14	518	0	15	445	0	16	211	0
14	526	0	15	462	0	16	214	0
14	527	0	15	462	0	16	215	0
14	527	0	15	468	0	16	222	0
14	539	0	15	469	0	16	223	0
14	539	0	15	470	0	16	223	0
14	539	0	15	471	0	16	225	0
14	540	0	15	471	0	16	228	0
14	540	0	15	475	0	16	232	0
14	546	0	15	477	0	16	236	0
14	551	0	15	483	0	16	236	−2
14	562	0	15	488	0	16	238	0
14	567	0	15	492	0	16	239	0
14	568	0	15	497	0	16	244	0
14	569	0	15	499	0	16	245	0
14	575	0	15	505	0	16	252	0
14	576	0	15	505	0	16	252	0
14	602	0	15	511	0	16	254	0
14	602	0	15	516	0	16	255	0
14	603	0	15	520	0	16	261	0
14	606	0	15	526	0	16	266	0
14	609	0	15	527	0	16	266	0
14	609	0	15	533	0	16	267	0
14	632	0	15	540	0	16	267	0
14	633	0	15	553	0	16	275	0
14	648	0	15	555	0	16	281	0
14	672	0	15	557	0	16	290	0
14	694	0	15	562	0	16	310	0
14	757	0	15	570	0	16	311	0

How these different types of observations are used in the analysis will be outlined in the respective sections of Chapter 5.

4.4 An artificial teaching example

In order to illustrate the different methods that will be given in the next chapter with a smaller example for which the computations required can be done by hand, a constructed example is given here. The hypothetical experiment comprises three groups of 15 animals; group 1 may be considered an undosed control group, group 2 a low-dose group, and group 3 a high-dose group. Time to death is given for each animal (scale unimportant), and the context of observation of each tumour is recorded as follows:

0 tumour not present,
1 incidental,
2 probably incidental,
3 probably fatal,
4 fatal.

The data are given in Table 4.4; a detailed solution to this example can be found in Appendix I.

Table 4.4 Time to death and context of observation for animals in three groups (hypothetical example)

Group 1		Group 2		Group 3	
Time	Context	Time	Context	Time	Context
140	2	100	2	80	0
140	0	110	0	90	4
140	0	110	0	100	0
150	0	120	0	110	0
150	0	130	3	130	2
160	0	140	2	140	1
165	2	150	2	140	3
170	4	150	0	140	0
180	0	160	4	150	0
185	4	165	0	150	4
200	1	170	0	150	1
200	1	180	3	160	4
200	0	200	1	170	0
200	0	200	0	190	3
200	0	200	1	200	0

5. NONPARAMETRIC METHODS FOR ANIMAL CARCINOGENESIS EXPERIMENTS

CHAPTER 5

NONPARAMETRIC METHODS FOR ANIMAL CARCINOGENESIS EXPERIMENTS

5.1 Introduction

This chapter examines the statistical evaluation of an animal carcinogenesis experiment with the goal of determining whether or not a test compound induces tumours. In most of the chapter, it is assumed that the animals were assigned to different exposure groups in a completely randomized design. Each group received a different dose level of a test compound or served as a control group, and the animals were examined for the presence of tumours either continuously (for observable tumours) or at necropsy (for occult tumours). As noted in Chapter 2, an evaluation of tumour occurrence data requires the examination of mortality patterns in the various groups. Accordingly, Section 5.3 describes the computation of nonparametric survival functions and nonparametric test statistics, which permit a comparison of mortality patterns among the different exposure groups; Section 5.4 describes methods for comparing the crude tumour rates of the different groups; Section 5.5 describes the method of Hoel and Walburg (1972) and other methods for nonfatal tumours; Section 5.6 describes the use of failure-time methods to analyse tumour incidence data for observable tumours or tumour mortality data for rapidly lethal tumours; and Section 5.7 discusses the method of Peto for analysing tumour data in which the context of observation of each tumour is known and tumours are observed in both the fatal and incidental contexts (Peto, 1974; Peto *et al.*, 1980). Many of the nonparametric test statistics presented in Sections 5.3 to 5.7 are closely related in functional form. Thus, Section 5.2 presents technical details common to the computation of nonparametric test statistics discussed in Sections 5.3–5.7.

5.2 Computation of nonparametric test statistics

Suppose that the animals have been randomized into $I + 1$ experimental groups, and that the animals in the ith group are exposed to a dose level d_i of a test compound, with $d_0 < d_1 < \cdots < d_I$, for $i = 0, 1, \ldots, I$. Often, the group indexed by 0 will be a control group, with $d_0 = 0$. Suppose that observations of the experimental endpoint of interest (e.g., death or occurrence of a tumour) are made at K distinct times $t_k, k = 1, 2, \ldots, K$. The data corresponding to each experimental endpoint may be summarized in K $2 \times (I + 1)$ contingency tables ($K \geq 1$). The kth contingency table takes the form of Table 5.1, where x_{ik} denotes the number of events (e.g., deaths or

Table 5.1 Summary of data corresponding to a particular experimental endpoint for all animals in risk set k

Dose level	d_0	d_1	$\cdots$	d_i	$\cdots$	d_I	Total
No. of events	x_{0k}	x_{1k}	$\cdots$	x_{ik}	$\cdots$	x_{Ik}	$x_{.k}$
No. of animals at risk	n_{0k}	n_{1k}	$\cdots$	n_{ik}	$\cdots$	n_{Ik}	$n_{.k}$

animals with tumour) observed in the ith group at t_k, and n_{ik} denotes the number of animals at risk in group i, for $k = 1, 2, \ldots, K$. The $n_{.k}$ animals for which data are summarized in the kth table will be referred to as risk set k. The definition of risk set will vary according to the experimental situation being considered.

The expected number of events in the ith exposure group for risk set k using indirect standardization is $E_{ik} = x_{.k}A_{ik}$, where $A_{ik} = n_{ik}/n_{.k}$. Thus, the observed and expected number of events in the ith group over the entire experiment are $O_i = \sum_{k=1}^{K} x_{ik}$ and $E_i = \sum_{k=1}^{K} E_{ik}$, respectively, for $i = 0, 1, \ldots, I$. Define

$$D_i = O_i - E_i = \sum_{k=1}^{K} (x_{ik} - E_{ik}) \tag{5.1}$$

and

$$V_{hi} = \sum_{k=1}^{K} \alpha_k A_{hk}(\delta_{hi} - A_{ik}) \tag{5.2}$$

where $\alpha_k = x_{.k}(n_{.k} - x_{.k})/(n_{.k} - 1)$ and δ_{hi} is defined as 1 if $h = i$ and 0 otherwise, for $h, i = 0, 1, \ldots, I$. Then, letting $\mathbf{D}' = (O_0 - E_0, O_1 - E_1, \ldots, O_I - E_I)$ be the vector of deviations of expected from observed values, and letting $\mathbf{V}$ be the $(I+1) \times (I+1)$ matrix with $(h+1, i+1)$ entry V_{hi}, a statistic to test for heterogeneity among the $I+1$ groups with respect to the rate of occurrence of the experimental endpoint in question may be calculated as

$$X_H^2 = \mathbf{D}'\mathbf{V}^-\mathbf{D}, \tag{5.3}$$

where $\mathbf{V}^-$ is a generalized inverse of $\mathbf{V}$. The statistic X_H^2 may be computed as $X_H^2 = \mathbf{D}_1'\mathbf{V}_1^{-1}\mathbf{D}_1$, where $\mathbf{D}_1$ is the vector of dimension I obtained by deleting $O_0 - E_0$ from the vector $\mathbf{D}$, and $\mathbf{V}_1$ is the $I \times I$ matrix of full rank obtained by deleting the first row and column of the matrix $\mathbf{V}$. If there is no difference among exposure groups with respect to the distribution of the occurrence of the endpoint in question, then X_H^2 will have an asymptotic chi-squared distribution with I degrees of freedom. A one-degree-of-freedom chi-squared test for an increasing or decreasing rate of occurrence of the endpoint in question with increasing dose level can be calculated as

$$X_T^2 = (\mathbf{d}'\mathbf{D})^2/(\mathbf{d}'\mathbf{V}\mathbf{d}), \tag{5.4}$$

where $\mathbf{d}' = (d_0, d_1, \ldots, d_I)$ is the vector of dose levels, and a test for departure from a monotone dose-response relationship can be based on

$$X_Q^2 = X_H^2 - X_T^2, \tag{5.5}$$

which has a chi-squared distribution with $I-1$ degrees of freedom under the null hypothesis that the dose-response relationship is linear.

In computing the above statistics, the deviations of expected from observed values from different risk sets are given equal weight. It is sometimes of interest to weight certain risk sets more heavily than others. Accordingly, define

$$D_{iW} = \sum_{k=1}^{K} w_k(x_{ik} - E_{ik}) \tag{5.6}$$

and

$$V_{hiW} = \sum_{k=1}^{K} w_k^2 \alpha_k A_{hk}(\delta_{hi} - A_{ik}), \tag{5.7}$$

where α_k, A_{ik}, E_{ik}, and δ_{hi} are defined above, and the w_k are non-negative weights. Then, letting $\mathbf{D}_W'$ denote the vector $(D_{0W}, D_{1W}, \ldots, D_{IW})$ and $\mathbf{V}_W$ denote the $(I+1) \times (I+1)$ matrix with $(h+1, i+1)$ entry V_{hiW}, a weighted statistic to test for heterogeneity among the $I+1$ groups with respect to the rate of occurrence of the experimental endpoint in question may be calculated as

$$X_{WH}^2 = \mathbf{D}_W' \mathbf{V}_W^- \mathbf{D}_W, \tag{5.8}$$

where $\mathbf{V}_W^-$ is a generalized inverse of $\mathbf{V}_W$. The statistic X_{WH}^2 may be computed as $X_{WH}^2 = \mathbf{D}_{W1}' \mathbf{V}_{W1}^{-1} \mathbf{D}_{W1}$, where $\mathbf{D}_{W1}$ is obtained by deleting D_{0W} from $\mathbf{D}_W$ and $\mathbf{V}_{W1}$ is obtained by deleting the first row and column of $\mathbf{V}_W$. If there is no difference among exposure groups with respect to the distribution of the occurrence of the endpoint in question, and if the weights are chosen properly, then X_{WH}^2 will have an asymptotic chi-squared distribution with I degrees of freedom. A one-degree-of-freedom test for an increasing or decreasing rate of occurrence of the endpoint in question with increasing dose level can be calculated as

$$X_{WT}^2 = (\mathbf{d}' \mathbf{D}_W)^2 / (\mathbf{d}' \mathbf{V}_W \mathbf{d}), \tag{5.9}$$

and a test for departure from a monotone dose-response relationship can be based on

$$X_{WQ}^2 = X_{WH}^2 - X_{WT}^2, \tag{5.10}$$

which has a chi-squared distribution with $I-1$ degrees of freedom under the null hypothesis that the dose-response relationship is linear.

In discussing X_H^2 and X_{WH}^2 above, it is stated that these statistics have asymptotic chi-squared distributions with I degrees of freedom under the null hypothesis. Exceptions may occur in the analysis of tumour data from experiments in which some groups have extremely high early mortality rates (e.g., due to toxicity of the test compound). For example, if all animals in the ith group die prior to observation of the first tumour in all $I+1$ groups, then $O_i = E_i = 0$, and the ith group makes no contribution to the above test statistics. In such cases, under the null hypothesis of homogeneity among groups, X_H^2 and X_{WH}^2 will have asymptotic chi-squared distributions with $I-r$ degrees of freedom, where r denotes the number of groups for which $O_i = E_i = 0$. Similarly, if the response is linear, X_Q^2 and X_{WQ}^2 will have asymptotic chi-squared distributions with $I-r-1$ degrees of freedom. The computation of the test statistics proceeds exactly as above, using only data from those groups for which

$E_i > 0$. When in all risk sets the n's are nearly zero for a particular dose, i.e., there is (almost) no animal at risk in that group, the asymptotic distributions may then not be valid. Thus, dose groups for which E_i is near zero should be omitted in calculating the test statistics, with a corresponding reduction in degrees of freedom for the tests of heterogeneity and of departures from a monotone dose-response relationship.

When results are available from several strata (see Section 2.5), an analysis combining the evidence from all strata can easily be obtained using the statistics described in this section. Let the strata be indexed by j, for $j = 1, \ldots, J$. Restricting the above methods to data from stratum j yields a vector of weighted observed minus expected, $\mathbf{D}_{Wj}$, and an associated covariance matrix, $\mathbf{V}_{Wj}$. Then, the combined analysis proceeds exactly as described in equations (5.8), (5.9) and (5.10), with $\mathbf{D}_W$ replaced by $\sum_{j=1}^{J} \gamma_j \mathbf{D}_{Wj}$, and $\mathbf{V}_W$ replaced by $\sum_{j=1}^{J} \gamma_j^2 \mathbf{V}_{Wj}$, where the γ_j depend on the choice of weights, w_k, in (5.6).

5.3 Nonparametric analysis of survival curves

The first step in evaluating an animal carcinogenesis experiment is to determine the effect of exposure to the test substance on mortality. Suppose that deaths are observed at K distinct times t_k, $k = 1, 2, \ldots, K$. For the purposes of summarizing the effect of exposure to the test compound on mortality, the times of death for animals killed accidentally or in planned sacrifices are considered to be censored observations. Animals lost to observation are considered censored at the time they were last under observation. The mortality data at time t_k may be summarized as in Table 5.1, where x_{ik} is the number of deaths in group i at time t_k, and n_{ik} is the number of animals in group i at risk of dying at t_k (i.e., the number of animals that die at or after t_k). For any time t, let $R(t) = \{k : t_k \leq t\}$; that is, $R(t)$ is the set of all k with index times of deaths occurring at or before t. Then, the Kaplan–Meier estimator of the survival function for group i is the step function defined as (Kaplan & Meier, 1958)

$$\hat{S}_i(t) = \prod_{k \in R(t)} \left(1 - \frac{x_{ik}}{n_{ik}}\right),$$

and the variance of $\hat{S}_i(t)$ may be estimated by

$$V\{\hat{S}_i(t)\} = \hat{S}_i^2(t) \sum_{k \in R(t)} \frac{x_{ik}}{n_{ik}(n_{ik} - x_{ik})}.$$

A plot of the $(I + 1)$ estimators, $\hat{S}_i(t)$ from the beginning of the experiment until terminal sacrifice reveals any effects of exposure to the test compound on mortality. Nonparametric estimates of percentiles can be obtained from the Kaplan–Meier survival curve (Miller, 1981b, pp. 74–75), and corresponding confidence intervals can be calculated. (See Slud et al., 1984, for a review and comparison of several available methods.)

Consider the data presented in Table 4.1 from a bioassay of 1,2-dichlorethane using female B6C3F1 mice. All deaths were due to natural causes except for one accidental death at 22 weeks in the control group and the 69 deaths at terminal sacrifice after 90 weeks on study. The Kaplan–Meier survival curves for the control, low-dose and

Fig. 5.1   Kaplan–Meier estimates of survival curves for three groups of female mice (□, control; ○, low dose; ▲, high dose) treated with 1,2-dichloroethane

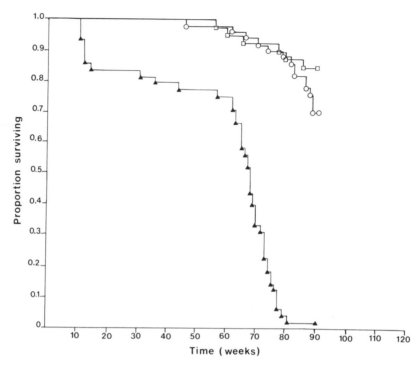

high-dose groups are given in Figure 5.1. Although there was only a slight increase in mortality in the low-dose group compared to the control group, there was a substantial increase in mortality in the high-dose group. Thus, it is clear that in comparing the proportions of animals with tumour in the high-dose group to those in the control or low-dose group, some consideration must be given to the possibility of bias due to the greater mortality in the high-dose group.

It may not always be necessary to test formally for differences in mortality patterns, as any difference in survival can lead to some degree of bias in the comparison of tumour rates and should, irrespective of its significance, be adjusted for. Nonetheless, formal comparisons can easily be made using generalized rank tests for censored data. The most widely used statistic for testing for survival differences is the generalized Savage statistic, often referred to as the log-rank statistic (Mantel, 1966; Cox, 1972), which is computed using (5.3). The corresponding trend statistic X_T^2, computed using (5.4), and departure from trend statistic X_Q^2, computed using (5.5), were presented by Tarone (1975).

For the data on female mice treated with 1,2-dichloroethane summarized in Figure 5.1, $X_H^2 = 127.8$, $X_T^2 = 85.3$ and $X_Q^2 = 42.5$, with degrees of freedom 2, 1 and 1, respectively. Thus, administration of 1,2-dichloroethane is clearly associated with increased mortality; however, the relationship is not strictly monotone in dose. The significance of X_Q^2 is due to the poor survival in the high-dose group relative to that in the controls, while the survival in the low-dose and control groups is similar (comparison of the low-dose and control groups yields $p = 0.21$).

The Wilcoxon rank sum test has also been modified for censored survival data, with two proposed modifications, both based on statistics that can be computed using (5.8). For the modified Wilcoxon test of Breslow (1970), w_k is taken to be $n_{.k}$; while for the modified Wilcoxon test of Peto and Peto (1972) and Prentice (1978), w_k is taken to be $\tilde{S}(t_k)$, where $\tilde{S}(t_k)$ is an estimator of the survival function calculated from the pooled data of all $I + 1$ groups. The corresponding trend test statistic X^2_{WT}, computed using (5.9), and departure from trend statistic X^2_{WQ}, computed using (5.10), were presented by Tarone and Ware (1977) and by Thomas et al. (1977). The modified Wilcoxon statistics are more sensitive than the generalized Savage statistics to differences in survival occurring early in an experiment, when a greater number of animals are at risk (Tarone & Ware, 1977; Thomas et al., 1977). For data from experiments with heavy interim sacrifices, the Peto–Prentice-modified Wilcoxon statistic should be used (Prentice & Marek, 1979).

For the data on 1,2-dichloroethane, the Breslow-modified Wilcoxon statistics are $X^2_W = 110.3$, $X^2_{WT} = 77.2$ and $X^2_{WQ} = 33.1$, with degrees of freedom 2, 1 and 1, respectively. The Wilcoxon statistics give slightly lower values than the corresponding Savage statistics because differences in survival are more pronounced at the end of the experiment.

In combining results of the above linear rank tests from several strata as suggested in Section 5.2, the appropriate stratum weights are $\gamma_j = 1$ for all j for the log-rank statistic and the Peto–Prentice-modified Wilcoxon statistic, and $\gamma_j = (N_j + 1)^{-1}$ for the Breslow-modified Wilcoxon statistic, where N_j is the total sample size in stratum j.

5.4 Analysis of crude proportions

In a well-designed and executed experiment in which there is no great disproportion in survival among the groups, one can usually obtain a good first indication of the possible significance of the results from analysis of the crude proportions of animals with tumour (Gart et al., 1979). This is, of course, the method traditionally used by toxicologists and pathologists. Although disproportionate survival may lead one astray, and such analyses are insensitive to differences in distributions of tumour occurrence or observation times, they are an instructive starting point for discussing the statistical analysis of tumour data.

Choice of denominator

The proportion consists of a numerator of the number of animals with the tumour of a specific site and/or type divided by a denominator of the number of animals at risk for that tumour. We consider three alternative ways of choosing the denominator:

(1) The number of animals initially put on test in each group.

(2) The number initially on test, less the numbers of animals which were not subjected to necropsy or for which the organ site in question was not submitted to or yielded tissue slides unsuitable for pathological examination. Thus, for instance, those

animals whose lungs were not available for pathological examination would be excluded from the denominators (and numerators) in the analyses of lung tumours, but if their livers were so examined, they would be included in the denominators of the analyses of liver tumours. In such cases, the denominators may vary from tumour site to tumour site within the same experiment.

(3) The initial number at risk, less those animals not subjected to necropsy or for which tissue slides for the organ in question are missing, and less, also, those animals 'dying early'. 'Dying early' may be defined in at least two ways:

(3a) Those dying before a pre-specified time on test, say, one year, before which time tumours almost never appear. This usually can be used only if the experimenter has reliable prior knowledge of the test animal and the tumour site. Of course, if a tumour is found before this time, one should use the next option.

(3b) Those dying before the first tumour at a specific site is found in any of the groups being compared. This again can lead to differing denominators for the various tumour sites within the same experiment.

The unadjusted denominator, although often used, is based on the tacit assumption that none of the missing animals had the tumour. One could also assume that all the missing animals had the tumour and adjust the numerators, rather than the denominators, accordingly. In some cases, these extreme possibilities for accounting for missing animals are analysed to determine if such extraordinary results could change the interpretation of the experiment. Such analyses have some polemic value but they are not usually presented in scientific publications. Therefore, we consider only the last two alternatives.

Alternatives (2) and (3) imply that the missing animals or those dying early are as likely to have had the tumour during their full lifetime as those that survived to terminal sacrifice and underwent a necropsy. The net effect of their deletion is to reduce the sample size, perhaps differentially among the groups. The typical outcome, particularly for the 'early death' correction, is that there is more early mortality among the higher-dose groups so that their denominators are reduced more than those of the control or lower-dose groups. Thus, although none of the numerators are changed, the proportions in the higher-dose groups are increased proportionately more. This may lead to a statistically significant positive dose-response or may erase an otherwise negative or inverse dose relation.

To illustrate these concepts and introduce some notation, consider again the data on 1,2-dichloroethane in Table 4.1, with lung as the target site. The two exposed groups consisted initially of 50 female animals, the design specifying equally spaced doses. The control group consisted initially of 40 female animals. The tumour under consideration (alveolar/bronchiolar adenoma) was first found in a high-dose animal dying at 62 weeks. The data may be summarized according to the various criteria in Table 5.2.

As the comparison of crude rates does not involve consideration of the time axis, we have simplified the notation for this Section 5.4 and use only the index for group $(i = 0, 1, \ldots, I)$ and suppress the time index $(k = 1, \ldots, K)$, which had been introduced in the general notation in Section 5.2.

Regardless of the choice of denominator, the statistical analyses may change in character depending on whether two or more than two groups are being compared and

Table 5.2 Number of animals with lung tumour and at risk (using different criteria) from data on 1,2-dichloroethane

Coded doses	$d_0 = 0$	$d_1 = 1$	$d_2 = 2$
No. with tumour	$y_0 = 2$	$y_1 = 7$	$y_2 = 15$
No. initially at risk	$n_0 = 40$	$n_1 = 50$	$n_2 = 50$
No. missing	0	0	2
No. at risk (Criterion 2)	$n_0' = 40$	$n_1' = 50$	$n_2' = 48$
No. dying before 62 weeks	3	2	12
No. at risk (Criterion 3b)	$m_0 = 37$	$m_1 = 48$	$m_2 = 36$

also on whether a large sample approximation may be legitimately employed rather than an exact or conditional analysis. We consider these cases in turn.

Comparison of two groups

Usually, this involves the comparison of a single-dose group and a control group. Consider the control group and the high-dose group in the data on 1,2-dichloroethane and lung tumours just presented. We use the denominator from the early-death criterion (3b). The notation and data are given in 2×2 tables in Table 5.3.

Table 5.3 2×2 table for comparison of control group and high-dose group from data on 1,2-dichloroethane

	Notation			Data		
Dose	d_0	d_2	Total	0	2	Total
Animals with tumour	y_0	y_2	s	2	15	17
Animals without tumour	$m_0 - y_0$	$m_2 - y_2$	$m_. - s$	35	21	56
	m_0	m_2	$m_.$	37	36	73

Approximate analyses of two groups

In our example, we consider the number of animals with tumour in the high-dose group as the observed quantity, $O = y_2$. Define the expected numbers of animals with tumour under the null hypothesis of no difference in tumour rates to be $E = (sm_2)/m_.$ for the high-dose group and $E' = (sm_0)/m_.$ for the control group. Define $D = O - E$. The variance of D under the null hypothesis is estimated by

$$V = \{s(m_. - s)m_0 m_2\}/\{m_.^2(m_. - 1)\}.$$

Alternatively, the reciprocal of this variance may be computed from the table of expected values,

$$1/V = \{(m_. - 1)/m_.\}\{1/E + 1/(m_2 - E) + 1/E' + 1/(m_0 - E')\}.$$

Many authors (Armitage, 1971, pp. 129 ff. and Snedecor & Cochran, 1980, pp. 124 ff.) use $m_.$ in place of $m_. - 1$ in the formula for V. Test statistics based on the above variance, however, have distributions better approximated by the normal or chi-square distribution in the unconditional sample space (Upton, 1982). The present formula is also better if one combines analyses for differing sexes and/or strains of test animals.

The approximately normal deviate test for equality of proportions is then

$$Z = D/\sqrt{V}.$$

Large positive values of Z indicate a direct or positive relation with the application of the compound and tumour production, and large negative values indicate an inverse or negative relation between application of compound and tumour production. One-tailed p-values are then read from tables of the normal or Gaussian distribution. Two-tailed tests are conveniently performed by considering

$$X^2 = Z^2 = D^2\{(m_{.} - 1)/m_{.}\}\{1/E + 1/(m_2 - E) + 1/E' + 1/(m_0 - E')\},$$

which is an approximate chi-square variate with one degree of freedom.

If a continuity correction is used, we have

$$Z_c = (D \pm \tfrac{1}{2})/\sqrt{V},$$

where $-\tfrac{1}{2}$ is used for a one-tailed test of a positive or direct relationship and $+\tfrac{1}{2}$ is used for a one-tailed test of a negative or inverse relation. Note that the continuity correction is employed to make the p-value for the approximate test closer to that of the exact or conditional test, which we discuss later. It has little effect for large numbers.

The significance test depends not only on the relative magnitude of the differential effect of the exposure on tumour production in the two groups but also on sample size. A commonly-used measure of this effect, which does not depend on sample size, is the odds ratio. This is the ratio of the odds of a tumour in the treated group to the corresponding odds in the control group. In the notation of early-death criterion (3b) this is estimated by the cross-product ratio:

$$\hat{R} = \{y_2.(m_0 - y_0)\}/\{y_0(m_2 - y_2)\},$$

where $\hat{R} > 1$ indicates a positive relation, $\hat{R} = 1$, no relation, and $\hat{R} < 1$, a negative or inverse relation of exposure with tumour production. Approximate confidence limits for this parameter can be computed by the method of Cornfield, which has been implemented in several computer programs (e.g., Thomas, 1975).

The question arises as to how large the numbers have to be to apply these approximate methods. The validity of these methods is not determined by the magnitude of the observed numbers themselves, but by the magnitude of the minimum of the expected values corresponding to the particular test or confidence interval method used. Thus, if

$$\min(E, m_2 - E, E', m_0 - E') \geq 1,$$

the Z_c-test should give a good approximation to the exact p-values. The accuracy of the approximate confidence interval also depends on the minimum expected values consistent with the marginal totals and the values of the odds ratios computed at the two confidence limits (see, e.g., Gart & Thomas, 1972). Thus, for instance, an experiment may be large enough to use approximate methods for a p-value but not for an upper 95% confidence limit.

Returning to our example, we have $O = 15$, and the table of expected values is shown in Table 5.4.

Table 5.4 Expected values in comparison of control group and high-dose group from data on 1,2-dichloroethane

	Dose		Total
	d_0	d_2	
With tumour	$E' = 8.62$	$E = 8.38$	$s = 17$
Without tumour	$m_0 - E' = 28.38$	$m_2 - E = 27.62$	$m_. - s = 56$
Total	$m_0 = 37$	$m_2 = 36$	$m_. = 73$

The minimum of the expected values, 8.38, is clearly large enough to apply the approximate test. Furthermore, we have

$$V = \{(17)(56)(37)(36)\}/\{(73)^2(72)\} = 3.3049,$$

or, alternatively,

$$1/V = (72/73)\{1/(8.38) + 1/(27.62) + 1/(8.62) + 1/(28.38)\} = 0.3026.$$

Thus,

$$Z = 6.62/\sqrt{(3.3049)} = 6.62\sqrt{(0.3026)} = 3.64,$$

for which the corresponding one-tailed $p = 0.00014$, indicating a highly significant positive difference between the high-dose and the control group. The two-tailed chi-square test yields

$$X^2 = Z^2 = 13.26, \qquad p = 0.00028.$$

The corresponding continuity corrected test is

$$Z_c = (6.12)/\sqrt{(3.3049)} = 3.37, \qquad p = 0.00038.$$

The cross-product ratio is $\hat{R} = \{(15)(35)\}/\{2(21)\} = 12.50$. The associated approximate 95% limits are $(2.35, 88.28)$. Checking the validity of the approximation at the limits, we compute the expected values in the four-fold tables with fixed marginals for the lower and upper limits. These are given in Table 5.5.

Table 5.5 Expected values in tables corresponding to lower and upper confidence limits

Lower limit:	$R_l = 2.35$	
5.8998	11.1002	17
31.1002	24.8998	56
37	36	
Upper limit:	$R_u = 88.28$	
0.3569	16.6431	17
36.6431	19.3569	56
37	36	

In the table corresponding to the lower limit, the minimal entry 5.8998 is greater than 1, whereas in the table corresponding to the upper limit, the minimal entry 0.3569 is less than 1. Thus, although the significance test and the lower 95% limit would appear to be approximately correct, the upper limit 95% cannot be relied upon here.

If $m_.$ is substituted for $m_. - 1$ in V, then the Z_c test will always agree with the 95% confidence interval in excluding the odds ratio of $R = 1$ whenever the one-tailed p-value is less than 0.025 and *vice versa*. It will usually agree, as in this case, with the computation of Z_c.

Exact or conditional analyses of two groups

When the numbers are small, exact or conditional analyses are feasible and may be necessary. The theoretical basis of such analyses is the initial randomization of the animals into two groups (Gart *et al.*, 1979). Consider the 73 animals in our example to be randomly divided into two groups of 37 and 36. If exposure to the chemical does not change the risk of tumour, then, regardless of the outcome of the randomization, 17 animals are fated to have this tumour. This 'fixing' of 17 as the marginal total is the reason for calling this analysis 'conditional'. Now consider the actual outcomes of the experiment, i.e., in this particular randomization, $y_2 = 15$, and those possible outcomes 'more extreme' in the positive direction, in this case $y_2 = 16$, and $y_2 = 17$. It is a simple combinatorial exercise to count the numbers of ways in which these outcomes can occur relative to the total number of possible randomizations. The ratio of these numbers is the precise one-tailed p-value for the Fisher–Irwin exact test.

We put this argument in mathematical notation. The total number of possible randomizations of $m_.$ animals having s tumours is given by the binomial coefficient $\binom{m_.}{s}$. For any integers u and v, for which $0 \leq u \leq v$, $\binom{v}{u}$ is also referred to as the number of ways of choosing u objects from v objects, is given by

$$\binom{v}{u} = \frac{v(v-1)\cdots(v-u+1)}{1\cdot 2\cdots u}.$$

The number of ways in which there can be y animals with tumour in the dose group and $s - y$ in the control group is

$$\binom{m_0}{s-y}\binom{m_2}{y}, \qquad y = 0, 1, \ldots, s.$$

The conditional probability that y occurs is thus

$$P(y \mid s) = \binom{m_0}{s-y}\binom{m_2}{y} \Big/ \binom{m_.}{s}, \qquad y = 0, 1, \ldots, s.$$

The exact one-tailed p-value for a possible increase in tumour incidence in the dose group is then

$$p = \sum_{y=y_2}^{s} P(y \mid s).$$

In applying this formula, note that $\binom{k}{j}$ is defined as zero when $j > k$.

Two-tailed exact tests are not so simply defined, since 'more extreme' does not have a unique meaning for unequal sample sizes. A reasonable procedure is to define p by accumulating all y such that $P(y \mid s) \leq P(y_2 \mid s)$, where y_2 is the observed outcome. When the sample sizes are equal, this rule leads to a p-value simply twice that for a one-tailed test. Otherwise, one may use the special tables of Armsen (1955).

Conditional point estimates of the odds ratio and exact confidence limits for this parameter have been described by Fisher (1935), Cornfield (1956) and Gart (1970), and their computation usually requires a computer program (see, e.g., Thomas, 1975). Programs that have been developed for pocket calculators can also be used (Rothman & Boice, 1979).

Consider again our example. We find

$$P(15 \mid 17) = \binom{37}{2}\binom{36}{15} \Big/ \binom{73}{17} = 0.00021$$

$$P(16 \mid 17) = \binom{37}{1}\binom{36}{16} \Big/ \binom{73}{17} = 0.00002$$

$$P(17 \mid 17) = \binom{37}{0}\binom{36}{17} \Big/ \binom{73}{17} = 0.00000$$

and thus $p = 0.00023$.

Recall that the approximate one-tailed Z_c test yielded a comparable value of $p = 0.00038$.

The computer program of Thomas (1975) yields the conditional maximum likelihood estimator for the odds ratio of 12.08 *versus* the cross-product ratio of 12.50 noted previously. Similarly, the exact 95% confidence limits for the odds ratio are (2.44, 119.33). The lower limit is comparable to the approximate value, 2.35, but the upper limit is quite different from the approximate upper limit, 88.28. This confirms the previous finding that the approximate upper limit is not reliable because it depends on a very small expected value. Note also that the exact limits include the null value of 1 whenever the appropriate one-tailed exact p is greater than 0.025, and will exclude 1 when p is less than 0.025.

Comparison of several groups

The usual design has one control group and at least two dose groups of a compound under test. One is usually interested in testing whether the proportion of animals with tumour increases or decreases monotonically with dose; that is, if $p(d_i)$, $i = 0, \ldots, I$ are the true proportions of the tumour among the various groups, whether $p(d_i)$ is a monotonic function of dose. A convenient monotonic function is the logistic,

$$p(d_i) = \exp(\alpha + \beta d_i)/\{1 + \exp(\alpha + \beta d_i)\}, \qquad i = 0, 1, 2, \ldots, I,$$

where, typically, $d_0 = 0$, corresponding to the control group. This may be written in the

logarithmic scale as
$$\log\{p(d_i)/q(d_i)\} = \alpha + \beta d_i,$$
where $q(d_i) = 1 - p(d_i)$. This implies that the log odds (or logit) is a linear function of dose. Alternatively, this means that the odds ratio between two doses d_i and d_j is
$$R_{ij} = \{p(d_i)q(d_j)\}/\{q(d_i)p(d_j)\} = \exp\{\beta(d_i - d_j)\}.$$
Thus, if the doses are equally spaced, say at unit intervals, the model implies that odds ratios between adjacent doses are equal. This model has properties that enable the extension of simpler exact tests to more complex situations (Cox, 1958, 1970, Chapter 5). It should be pointed out, however, that most of the tests based on the logistic model are robust, that is, they are valid regardless of whether this model holds exactly, and many can also be justified from completely model-free considerations.

The data are usually arrayed in a $2 \times (I + 1)$ table, as in Table 5.6.

Table 5.6   Notation for data from experiment with $I + 1$ groups

	Dose				Total
	d_0	d_1	$\cdots$	d_I	
With tumour	y_0	y_1	$\cdots$	y_I	s
Without tumour	$m_0 - y_0$	$m_1 - y_1$	$\cdots$	$m_I - y_I$	$m_. - s$
Total	m_0	m_1	$\cdots$	m_I	$m_.$

Our numerical example, with the elimination of the early-death criterion (3b), is given in Table 5.7.

Table 5.7   Data on lung tumour for three groups from study on 1,2-dichloroethane

	Dose			Total
	0	1	2	
With tumour	2	7	15	24
Without tumour	35	41	21	97
Total	37	48	36	121

Approximate analyses of several groups

Let us denote the observed numbers with tumour as $O_i = y_i$, $i = 0, 1, \ldots, I$, and their corresponding expected values under the null hypothesis of no difference as
$$E_i = (sm_i)/m_., \qquad i = 0, 1, \ldots, I,$$
where $m_. = \sum_{i=0}^{I} m_i$. Defining $D_i = O_i - E_i$, the test statistic for possible monotonic trend with dose is based on
$$T = \sum_{i=0}^{I} d_i D_i.$$

Under the null hypothesis, T has mean zero, and its variance is estimated by

$$V_T = [\{s(m_. - s)\}/\{m_.(m_. - 1)\}] \sum_{i=0}^{I} m_i(d_i - \bar{d})^2,$$

where $\bar{d} = (\sum_{i=0}^{I} m_i d_i)/m_.$. The Cochran–Armitage normal deviate test for trend (see, e.g., Armitage, 1971, pp. 363–365) is then

$$Z_T = T/\sqrt{V_T}.$$

If the doses are equally spaced, say, with interval Δ, a simple continuity correction may be easily employed to yield,

$$Z_{Tc} = (T \mp \Delta/2)/\sqrt{V_T},$$

where the minus or plus signs are used for one-tailed testing against a direct and inverse relation, respectively. In the case of unequally spaced doses, the continuity correction is less readily applied (Kendall & Stuart, 1961, p. 508). Although the actual level of the Cochran–Armitage trend test can deviate from the nominal level for asymmetric designs (Portier & Hoel, 1984b), this can be remedied by a Cornish–Fisher skewness correction (Tarone, 1986).

Two-tailed tests may be based on the squared value of Z_T, $X_T^2 = Z_T^2$, which is an approximate chi-square variate with one degree of freedom. For $I = 1$, these tests are exactly equivalent to the approximate tests for comparing two groups. Although the Cochran–Armitage test follows from the assumption of a logistic model, Tarone and Gart (1980) showed that it is asymptotically, locally fully efficient for testing the null hypothesis against any choice of a monotonic, locally linear function. The approximate tests should be adequate as long as the minimum expected value exceeds one.

The question arises whether a linear relation is an appropriate alternative. Testing for this possibility is facilitated by first computing the usual chi-square for heterogeneity in a $2 \times (I + 1)$ contingency table:

$$X_H^2 = \{(m_. - 1)/m_.\}\left[\sum_{i=0}^{I} D_i^2\{1/E_i + 1/(m_i - E_i)\}\right], \qquad (5.11)$$

which is approximately distributed as a chi-square variate with I degrees of freedom if all of the $p(d_i)$ are equal. This statistic can also be used for testing for heterogeneity among groups in which there is no quantitative dose relationship in the treatment regimens, i.e., differing chemicals and/or vehicles or other differing control groups. For the dose-relation situation, the approximate chi-square statistic with $I - 1$ degrees of freedom for departure from linear trend is $X_Q^2 = X_H^2 - X_T^2$.

This computation is usually presented in a table analogous to an analysis-of-variance table. If $I = 2$, the chi-square statistic X_Q^2 is the appropriate statistic for testing the possibility of a quadratic relation (hence, the subscript Q). When $I \geq 3$, it is the statistic for an omnibus test of all nonlinear polynomial coefficients, i.e., quadratic, cubic, quartic, etc. When $I = 1$, $X_T^2 \equiv X_H^2$ and thus $X_Q^2 \equiv 0$, and no test of departure is possible.

The strength of the possible effect of dose can be estimated by fitting the linear logistic model. We may employ the method of maximum likelihood (see, e.g., Thomas

& Gart, 1983) to estimate β by $\tilde{\beta}$. The odds ratio, R_i, between any dose d_i and the control $d_0 = 0$, is then estimated by $\tilde{R}_i = \exp(\tilde{\beta}d_i)$. Alternatively, a much more simply computed estimator of R_i is found from the cross-product ratio. These estimators are $\hat{R}_i = \{O_i(m_0 - O_0)\}/\{O_0(m_i - O_i)\}$, $i = 1, \ldots, I$. This appears to be a reasonably good estimator for $\frac{1}{3} \le R_i \le 3$, but is otherwise biased towards unity.

Returning to our example, our preliminary calculations are illustrated in Table 5.8.

Table 5.8 Expected values for data on 1,2-dichloroethane (Table 5.7)

		Dose group			Total
	d_i	0	1	2	
No. with tumour	O_i	2	7	15	24
Expected no.	E_i	7.34	9.52	7.14	24
	$m_i - E_i$	29.66	38.48	28.86	97
Total	m_i	37	48	36	121
	$O_i - E_i = D_i$	-5.34	-2.52	$+7.86$	0.00

As the minimum expected value exceeds one, the approximate tests should be adequate.

$$T = (-5.34)(0) + (-2.52)(1) + (7.86)(2) = 13.20,$$

and

$$\sum_{i=0}^{2} m_i(d_i - \bar{d})^2 = \sum_{i=0}^{2} m_i d_i^2 - \left(\sum_{i=0}^{2} m_i d_i\right)^2 \Big/ m.$$

$$= 192 - (120)^2/121 = 72.9917.$$

Therefore,

$$V_T = [\{(24)(97)\}/\{(121)(120)\}](72.9917) = 11.7028.$$

The one-tailed test for positive trend yields

$$Z_T = (13.20)/\sqrt{(11.7028)}$$
$$= 3.86, \qquad p = 0.00006,$$

and the two-tailed chi-square test is

$$X_T^2 = Z_T^2 = (3.86)^2 = 14.90, \qquad p = 0.00011.$$

The corresponding continuity corrected test is

$$Z_{Tc} = (12.70)/\sqrt{(11.7028)} = 3.71, \qquad p = 0.00010.$$

Turning to the question of possible departure from linearity, we compute $X_H^2 = 16.33$ according to (5.11) and derive Table 5.9. Clearly, there is no evidence that a linear model does not fit.

If we fit the logistic model by maximum likelihood, we find $\tilde{\beta} = 1.32 \pm 0.37$. The

Table 5.9 Summary of analysis of crude proportions for the 1,2-dichloroethane example

Source of variation	Degrees of freedom	Chi-square	p
Linear trend	1	$X_T^2 = 14.90$	0.00011
Departure from linearity	1	$X_Q^2 = 1.43$	0.23
Total (heterogeneity)	2	$X_H^2 = 16.33$	0.00028

estimates of the odds ratio of the dosed to control groups, using this value and the cross-product ratios, are given in Table 5.10. The good agreement between these estimates reflects the fact that the linear logistic model fits these data well.

Table 5.10 Estimates of odds ratio from data on 1,2-dichloroethane

	Dose group		
$d_i = i$	0	1	2
$\hat{R}_i$	1.00	3.60	12.50
$\hat{R}_i = \exp(\tilde{\beta}i)$	1.00	3.74	14.01

Exact or conditional analyses for linear trend

When the numbers are small, it may be necessary to use the exact test for trend (Cox, 1958), which is a generalization of the Fisher–Irwin test. This test statistic can be derived from the logistic model, but the null hypothesis and the distribution used to obtain a p-value hold very generally. Like the exact test for two groups, its theoretical basis is the randomization of the animals into several groups. Under the null hypothesis, it is assumed that the total number of animals with tumour in all the groups is fixed at s. The conditional distribution is then

$$P(y_0, y_1, \ldots, y_I \mid s) = \binom{m_0}{y_0}\binom{m_1}{y_1} \cdots \binom{m_I}{y_I} \bigg/ \binom{m.}{s}$$

where $\sum_{i=0}^{I} y_i = s$. The observed statistic for which probabilities of more extreme outcomes will be calculated is $\sum_{i=0}^{I} O_i d_i = A$. For tests of positive trend, the p value is computed from

$$p = \sum_{\Omega} P(y_0, y_1, \ldots, y_I \mid s).$$

where Ω consists of all possible values of $y_i \geq 0$ such that $\sum_{i=0}^{I} y_i = s$ and $\sum_{i=0}^{I} y_i d_i \geq A$. For a test of a negative trend, the sense of the last inequality is reversed. The application of this test usually requires a computer program (see, e.g., Thomas *et al.*, 1977). For $I = 1$, this test is identical to the exact test for two groups. Cox (1958) showed that the Cochran–Armitage test is the normal approximation to the exact randomization trend test.

It is also possible to perform an exact test for departure from linearity. This requires further conditioning on the observed value of $\sum y_i d_i$, which results in a more complex distribution. Bayer and Cox (1979) have published a program that can be used for this purpose. Conditional maximum likelihood methods can be used in small numbers to estimate β, and these have been implemented, in a somewhat more general context, by Smith *et al.* (1981).

Returning briefly to our example, we find that the conditional test for linearity, fixing $s = 24$, yields the exact one-tailed $p = 0.00007$, which is quite close to the value found from the continuity-corrected Z_{Tc}, specifically, $p = 0.00010$.

Combination of results over sexes, strains or experiments

The approximate tests for trend can easily be combined over sexes, strains or experiments. In each analysis the doses must be uncoded or be coded in the same way. The combined normal deviate test statistic is calculated simply by adding the numerators and the squares of the denominators of the individual statistics and dividing the summed numerators by the square root of the summed squared denominators. Its mathematical formula is $Z_T = \sum T / \sqrt{(\sum V_T)}$, where the summation is over the different subexperiments.

The continuity corrected normal deviate test is $Z_{Tc} = (\sum T \pm \Delta/2) / \sqrt{(\sum V_T)}$, where the doses are equally spaced, Δ units apart in all experiments, and the minus is used for testing a direct relation and the plus for an inverse relation. Note that the $\Delta/2$ corrections are *not* summed over the several experiments in combining the test statistics. This is the so-called Mantel–Haenszel procedure (Mantel & Haenszel, 1959; Mantel, 1963), which is the optimal procedure for testing the common slope, β, of a stratified logistic model. For $I = 1$ it essentially reduces to Cochran's (1954) test for the combination of 2×2 tables. Radhakrishna (1965) and Tarone and Gart (1980) showed the asymptotic efficiency of these combined tests to be robust (or insensitive) to modest departures from this logistic model.

Combined approximate tests for departure from linearity or for homogeneity of slopes from different experiments, and the maximum likelihood estimation of a common β over several experiments, usually require the use of a computer program (see, e.g., Thomas & Gart, 1983). Exact combined tests also usually require such programs (for $I = 1$, see, e.g., Thomas, 1975; for $I \geq 2$, see, e.g., Bayer & Cox, 1979), as does calculation of the conditional maximum likelihood estimate (for $I = 1$, see Thomas, 1975; for $I \geq 2$, see Smith *et al.*, 1981).

The various combined tests may not be appropriate if the relative effect of treatment varies greatly over the various strata or experiments being combined. Such variation for logistic models is measured by differences in the odds ratio for one-dose experiments or by differences in logistic slope for multiple-dose experiments. Statistical tests for the homogeneity of odds ratios are given by Breslow and Day (1980, pp. 142–146; see also Tarone, 1985) and tests for homogeneity of logistic slopes by Thomas and Gart (1983).

The issue of multiple comparisons among doses

Often, experimenters compare each of the I dosed groups in turn with the control group and report the results of these several tests in addition to the trend test. This raises the problem of multiple comparisons. If there is a strong direct relationship with dose, analyses of the results of the high dose and perhaps of some lower doses agree with the trend test in finding significance. A problem of interpretation develops when the lower-dose comparison is significant while the high-dose comparison and the trend test are not. In such cases, the chi-squares for homogeneity and departure from trend are usually large, if not significantly so. An adjustment in significance may be employed to allow for the possibility of finding significance by any one of two or more statistical tests. The simplest and most widely used correction employs the Bonferroni inequality (Miller, 1981a, pp. 6–10). If the desired overall significance level for the test of the chemical compound at I doses is α, the individual comparison of the ith dose to control is made at a significance level α_i, where $\sum_{i=1}^{I} \alpha_i = \alpha$. Alternatively, the observed p-value for comparison of the ith dose to control is multiplied by α/α_i. Because greater emphasis should be given to significance at the highest dose, α_I should be chosen to be larger than the remaining α_i, $i = 1, \ldots, I-1$.

The Bonferroni correction may be used to adjust multiple tests both in the previously considered survival analyses as well as in the following analysis of prevalent and rapidly lethal (or observable) tumour rates. Unless the target organ for a given test compound is known in advance, control of the overall experimental error rate is necessary. This more difficult question of multiple comparisons over organ sites is discussed in Section 7.2.

5.5 Prevalence analysis for nonlethal occult tumours

Hoel and Walburg (1972) pointed out the importance, when evaluating data on occult tumours, of making a distinction between those tumours that are lethal and those that are nonlethal. Nonlethal occult tumours are discovered at necropsy, either after terminal sacrifice or after an animal has died prior to terminal sacrifice because of illness unrelated to the presence of the tumour. In this section, tests for the equality of prevalence rates for nonlethal occult tumours are presented. An assumption underlying the derivation of these prevalence tests is that, at least with regard to the presence or absence of a nonlethal tumour, death is a random sampling mechanism. This is an extremely strong assumption, implying that a tumour-bearing animal is, in every way except for the presence of a tumour, as healthy as a tumour-free animal. Nonetheless, such statistical procedures can be useful in evaluating the carcinogenic potential of a test compound. In this section it is assumed that all tumours of a particular type observed in the carcinogenesis experiment under consideration are nonlethal.

Suppose the carcinogenesis experiment extends from time zero to time T, where T denotes the time at which the terminal sacrifice is scheduled. Now suppose that the interval $(0, T)$ is subdivided into $J - 1$ subintervals, $\mathscr{I}_j$, where $\mathscr{I}_j = (T_{j-1}, T_j]$ for $j = 1, 2, \ldots, J - 2$, and $\mathscr{I}_{J-1} = (T_{J-2}, T_{J-1})$, with $T_0 = 0$ and $T_{J-1} = T$. Then, let the number of animals dying in group i during subinterval $\mathscr{I}_j$ be denoted by N_{ij} and the

Table 5.11   Summary of tumour prevalence data for nonlethal tumours in interval $\mathscr{I}_j$

	Dose						Total
	d_0	d_1	$\cdots$	d_i	$\cdots$	d_I	
No. of animals with tumour	Y_{0j}	Y_{1j}	$\cdots$	Y_{ij}	$\cdots$	Y_{Ij}	$Y_{.j}$
No. of animals dying in $\mathscr{I}_j$	N_{0j}	N_{1j}	$\cdots$	N_{ij}	$\cdots$	N_{Ij}	$N_{.j}$

number of these animals in which a tumour is discovered at necropsy be denoted by Y_{ij}, $i = 0, 1, \ldots, I$. Then, for each subinterval $\mathscr{I}_j$, the tumour prevalence data may be summarized in a $2 \times (I + 1)$ table such as Table 5.11.

In addition, the tumour prevalence of the animals killed at terminal sacrifice can be summarized in a similar table, indexed by J, where N_{iJ} denotes the number of animals in group i surviving to terminal sacrifice, and Y_{iJ} denotes the number of these animals in which a tumour is found at necropsy. Note that, if there are any planned interim sacrifices, each time of such a sacrifice is treated as a distinct subinterval and contributes a separate $2 \times (I + 1)$ table of tumour prevalence data. Once the tumour prevalence data have been stratified into J strata, as described above, tests for equality of tumour prevalence rates can be derived using standard contingency table methods.

The prevalence test statistics can be computed using (5.3), (5.4) and (5.5), where D_i and V_{hi} are calculated as in (5.1) and (5.2), after substituting J for K, j for k, Y_{ij} for x_{ik} and N_{ij} for n_{ik}. Let X_{PH}^2 denote the test for equality of tumour prevalence rates in the $I + 1$ groups using (5.3) (Armitage, 1966), X_{PT}^2 denote the corresponding trend test statistic computed using (5.4) (Mantel, 1963), and X_{PQ}^2 denote the corresponding departure from trend statistic computed using (5.5).

Consider now the data given in Table 4.1 on alveolar/bronchiolar adenomas in the experiment with 1,2-dichloroethane in female mice. In the opinion of a pathologist involved in evaluating this experiment, it was extremely unlikely that any of these adenomas contributed to the deaths of tumour-bearing animals. This claim is supported by the fact that, in the two groups (control and low-dose) with good survival, alveolar/bronchiolar adenomas were found only in animals surviving until terminal sacrifice. To demonstrate the prevalence methods for nonlethal tumours, let us first assume that the 90-week experiment was divided (prior to evaluation of the data) into three subintervals, $(0, 52]$, $(53, 72]$ and $(73, 90)$, with terminal sacrifice planned at 90 weeks. No tumour was found in animals dying in the first 52 weeks, and, hence, the prevalence analysis is based on the 2×3 contingency tables presented in Table 5.12.

Applying the above methods, we find $X_{PH}^2 = 15.10$, $X_{PT}^2 = 13.51$ and $X_{PQ}^2 = 1.59$, with 2, 1 and 1 degrees of freedom, respectively. Thus, administration of 1,2-dichloroethane is associated with increased tumour prevalence, and the increase is clearly dose-related.

The method of subdividing the length of the experiment into subintervals warrants further discussion. In the above analysis it was assumed that subdivisions were chosen *a priori*, without reference to the data on tumour prevalence. Peto *et al.* (1980) suggest an adaptive interval selection method in which subintervals are determined by the tumour prevalence data. This method is illustrated using the tutorial example of Peto *et*

Table 5.12 Contingency tables for preselected time intervals for prevalence analysis of data on 1,2-dichloroethane

$\mathscr{I}_1 = (53, 72)$

d_i	0	1	2
Y_{i1}	0	0	8
$N_{i1} - Y_{i1}$	3	3	14

$\mathscr{I}_2 = (73, 90)$

Y_{i2}	0	0	6
$N_{i2} - Y_{i2}$	3	11	8

$\mathscr{I}_3$: terminal sacrifice

Y_{i3}	2	7	1
$N_{i3} - Y_{i3}$	31	28	0

al. (1980) (Table 5.13), and is then applied to the data on 1,2-dichloroethane and lung adenoma.

The first step in the adaptive interval selection method, shown in row 1 of Table 5.13, is to list, in increasing order, the times of death of all animals (pooling times of death from all $I + 1$ exposure groups) for which necropsies were performed. The list starts with the first time at which an animal died and was subjected to necropsy – week 50. The times of death for animals in which a tumour was found are underlined. In the case of ties (i.e., times at which some animals had tumours but others did not), the animals with tumours are listed first. The second row of the table contains asterisks which separate the times into what Peto et al. refer to as 'ad-hoc runs'. Each ad-hoc run is a sequence of consecutive underlined times followed by a sequence of consecutive times without underlining. The times before that at which the first animal with a tumour was found do not play any further role in the analysis, as the prevalence is zero for that period. The third row gives the proportion of underlined times in each run; that is, this row gives the estimated tumour prevalence within each time interval

Table 5.13 Calculation of subintervals for prevalence analysis using the adaptive method of Peto et al. (1980)

(1)	50	67	67	67	67	94	97	105	110	110	115	115	120	121	124	124	128	130	134	137	139
(2)	*					*		*			*			*					*		*
(3)		0.25				0.50		0.67			0.67			0.20					0.67		
(4)	*					*		*			*								*		*
(5)		0.25				0.50		0.67				0.375							0.67		
(6)	*					*		*											*		*
(7)		0.25				0.50							0.455						0.67		
(8)	*					*													*		*
(9)		0.25										0.462							0.67		

defined by a run. The remaining rows summarize the process of merging adjacent runs for which the estimated prevalence rates decrease with increasing time, and deleting asterisks separating merged runs. This process of 'pooling adjacent violators' is equivalent to maximum likelihood estimation of prevalence rates assuming nonde-creasing prevalence (Ayer *et al.*, 1955). For the data in the table, the first decrease in row 3 occurs between the fourth run (estimated prevalence, 0.67) and the fifth run (estimated prevalence, 0.20). Merging of these runs forms a new fourth interval with an estimated prevalence of 0.375 (row 5). The first decrease in row 5 occurs between the third run (estimated prevalence, 0.67) and the new fourth interval. Merging of the third run and the fourth interval results in a new third interval with an estimated prevalence 0.455 (row 7). The first decrease in row 7 occurs between the second run (estimated prevalence, 0.50) and the new third interval. Merging of the second run with the third interval results in three intervals with increasing estimated prevalence rates (row 9). Thus, the adaptive method gives a subdivision into three time intervals: the first subinterval consists of week 67, with an estimated prevalence of 0.25; the second subinterval consists of weeks 94–130, with an estimated prevalence of 0.462; and the third subinterval consists of weeks 134–139, with an estimated prevalence of 0.67.

For the data on the effects of 1,2-dichloroethane in female mice on adenoma incidence, the above adaptive method leads to a single interim subinterval from 62 weeks (when the first adenoma was found) to 88 weeks. Thus, the prevalence tests using the adaptive interval selection method are based on the 2×3 contingency tables in Table 5.14.

For these tables, $X^2_{PH} = 14.49$, $X^2_{PT} = 13.23$ and $X^2_{PQ} = 1.25$, with degrees of freedom 2, 1 and 1, respectively. These values are smaller, but quite similar, to the values obtained previously using the prevalence analysis after a-priori subdivision into three subintervals.

Whatever the method of interval selection, it is possible to find subintervals in which deaths are observed in only one exposure group. Tumours found in such subintervals will be ignored in calculating the prevalence test statistics. All such subintervals occurring after the first tumour has been observed may be merged with adjacent intervals containing deaths in additional exposure groups, although care should be

Table 5.14 Contingency tables after time partition by ad-hoc runs for prevalence analysis of data on 1,2-dichloroethane

$\mathcal{I}_1 = [62, 90)$	d_i	0	1	2
	Y_{i1}	0	0	14
	$N_{i1} - Y_{i1}$	4	13	21

$\mathcal{I}_2$: terminal sacrifice				
	Y_{i2}	2	7	1
	$N_{i2} - Y_{i2}$	31	28	0

taken not to merge intervals with widely disparate baseline prevalence rates. One instance in the adaptive interval selection method in which this situation will arise routinely is when the last animal dying prior to terminal sacrifice has a tumour but the penultimate animal dying has no tumour. In this case, the last subinterval will contain only the animal which died last. Accordingly, the last time of death should be included in the immediately preceding subinterval, provided this preceding subinterval includes deaths from another exposure group.

For both methods of interval selection described above, when applied to the data for 1,2-dichloroethane, the resulting prevalence method chi-squared statistics for the effect of exposure on tumour prevalence are smaller than the corresponding chi-squared statistics based on the crude tumour rates after eliminating animals dying prior to observation of the first tumour (Table 5.9). For an experiment the size of that with 1,2-dichloroethane, such a finding is not unusual. Regardless of the interval selection method, some efficiency may be lost because of the small number of animals dying in control (and sometimes low-dose) groups prior to terminal sacrifice. This is better illustrated by considering what would have happened if there had been no low-dose group in the 1,2-dichloroethane experiment. In order to compare the control group and the high-dose group, the prevalence test using the adaptive interval selection method is based on the 2×2 contingency tables in Table 5.15.

The prevalence test for equality of tumour rates gives $X_{PH}^2 = X_{PT}^2 = 6.23$. The analysis of crude tumour rates after eliminating animals dying prior to observation of the first tumour (see Section 5.4) gave an approximate chi-squared test statistic for equality of tumour rates of $X^2 = 13.26$. Even though survival in the high-dose group is quite poor, the simpler analysis of adjusted crude tumour rates gives a much more significant result than the prevalence analysis. This is because only four control animals died prior to terminal sacrifice when all but one of the high-dose animals died, while only one high-dose animal survived to terminal sacrifice when the majority of the

Table 5.15 Contingency tables for comparison of control group and high-dose group from data on 1,2-dichloroethane

$\mathcal{I}_1 = [62, 79)$

d_i	0	2
Y_{i1}	0	13
$N_{i1} - Y_{i1}$	3	21

$\mathcal{I}_2 = [80, 85)$

	0	1
Y_{i2}	0	1
$N_{i2} - Y_{i2}$	1	0

$\mathcal{I}_3$: terminal sacrifice

	2	1
Y_{i3}	2	1
$N_{i3} - Y_{i3}$	31	0

control animals contribute to the prevalence analysis. For larger experiments, the efficiency of the prevalence analysis should improve relative to the crude tumour rate analysis. It is important to note that the inefficiency of the interval prevalence method in this example is due to the large difference in intercurrent mortality rates. If intercurrent mortality rates are equal in all groups, then the interval prevalence method will, in general, be more efficient than the crude tumour rate analysis (McKnight, 1981).

Dinse and Lagakos (1983) proposed a logistic regression method for analysing nonlethal tumour data. Their method does not require selection of time intervals but, rather, makes use of the time of death of each animal. Dinse and Lagakos assume that the tumour prevalence rate at time t for animals in the group exposed to dose level d_i of the test compound is given by

$$P(d_i; t) = \exp\{\gamma(t) + \delta_i\}/[1 + \exp\{\gamma(t) + \delta_i\}], \tag{5.11}$$

where $\gamma(t) = \beta_0 + \beta_1 t + \beta_2 t^2 + \cdots + \beta_r t^r$. The carcinogenic potential of the test compound is assessed by testing the null hypothesis $H_0: \boldsymbol{\delta} = \mathbf{0}$, where $\boldsymbol{\delta}' = (\delta_0, \delta_1, \ldots, \delta_I)$ is a vector of group-specific parameters. The validity of the test of H_0 rests on the assumption that the prevalence function under H_0 can be represented adequately by the logistic function

$$P(t) = \exp\{\gamma(t)\}/[1 + \exp\{\gamma(t)\}]. \tag{5.12}$$

The polynomial $\gamma(t)$ will be referred to as the prevalence log-odds function. As in Section 5.3, denote the kth time at which animal deaths are observed by t_k, and let x_{ik} denote the number of deaths observed in group i at time t_k, $k = 1, \ldots, K$; $i = 0, 1, \ldots, I$. Similarly, let Y_{ik} denote the number of animals in which a tumour is found at necropsy among the x_{ik} animals from group i dying at t_k. Let $\hat{\gamma}(t) = \hat{\beta}_0 + \hat{\beta}_1 t + \hat{\beta}_2 t^2 + \cdots + \hat{\beta}_r t^r$, where $\hat{\boldsymbol{\beta}}$ denotes the maximum likelihood estimator of $\boldsymbol{\beta} = (\beta_0, \beta_1, \ldots, \beta_r)'$ under $H_0: \boldsymbol{\delta} = \mathbf{0}$. It follows that score tests of H_0 can be based on the $(I + 1)$ statistics

$$\hat{D}_i = \sum_{k=1}^{K} (Y_{ik} - x_{ik} \hat{P}_k),$$

where $\hat{P}_k = \exp\{\hat{\gamma}(t_k)\}/[1 + \exp\{\hat{\gamma}(t_k)\}]$. Note that $\hat{D}_i$ can be written as $O_i - \hat{E}_i$, where O_i is the observed number of animals in group i in which tumours were discovered, and $\hat{E}_i$ is the expected number calculated on the basis of the estimated polynomial prevalence log-odds function. The covariance matrix $\hat{\mathbf{V}}$ of the vector $\hat{\mathbf{D}} = (\hat{D}_0, \hat{D}_1, \ldots, \hat{D}_I)'$ can be obtained using standard score test methodology, and a test of $H_0: \boldsymbol{\delta} = \mathbf{0}$ can be based on

$$\hat{X}_{PH}^2 = \hat{\mathbf{D}}' \hat{\mathbf{V}}^- \hat{\mathbf{D}},$$

which will have an asymptotic chi-squared distribution with I degrees of freedom under H_0, provided the prevalence function under H_0 can be described by (5.12) with $\gamma(t)$ an r degree polynomial in time. Similarly, writing $\delta_i = \Delta d_i$ for all i, a test for monotone trend in response can be derived as a score test of $H_0: \Delta = 0$, which leads to the test statistic

$$\hat{X}^2_{PT} = (\mathbf{d}'\hat{\mathbf{D}})^2/(\mathbf{d}'\hat{\mathbf{V}}\mathbf{d}).$$

Provided the null prevalence function can be described by (5.12) with $\gamma(t)$ an r degree polynomial, the statistic $\hat{X}^2_{PT}$ will be asymptotically distributed, under H_0, as a chi-squared random variable with one degree of freedom.

Although the method of Dinse and Lagakos (1983) does not require selection of time intervals, an appropriate degree polynomial must be selected to estimate the prevalence function. The importance of the choice of r is illustrated by the data from the 1,2-dichloroethane experiment. Using the data from all three exposure groups and assuming a linear prevalence log-odds function (i.e., $r = 1$ in $\gamma(t)$), one finds that $\hat{X}^2_{PT} = 17.67$. For the same data, assuming a quadratic prevalence log-odds function (i.e., $r = 2$ in $\gamma(t)$), one finds that $\hat{X}^2_{PT} = 9.93$. Similarly, when comparing the high-dose group to the control group, deleting the data from the low-dose group, one finds $\hat{X}^2_{PT} = 17.44$ with a linear prevalence log-odds function and $\hat{X}^2_{PT} = 6.04$ with a quadratic prevalence log-odds function. It should be noted that the disparity in the results obtained with linear and quadratic prevalence log-odds functions in this example is due primarily to the large differences in intercurrent mortality rates among groups. In such a situation, different choices of interval can similarly lead to widely disparate results using analysis based on $\hat{X}^2_{PT}$. McKnight (1985) has noted that, in cases of extreme differences in intercurrent mortality rates, all methods of time adjustment eventually break down.

This example raises the important issue of the need for further research on methods for choosing the degree of the polynomial, $\gamma(t)$. Using all three exposure groups and fitting (5.12) to the 1,2-dichloroethane data, the model with parameters β_0, β_1 and β_2 provides a significantly better fit than the model with only parameters β_0 and β_1 ($p = 0.0023$). Thus, selection of the best fitting polynomial, in the absence of information on exposure level, would lead to selection of a quadratic prevalence log-odds function. Letting $\delta_i = \Delta d_i$ for all i and fitting (5.11) to the 1,2-dichloroethane data, the model based on β_0, β_1, β_2 and Δ provides little improvement in fit over the model based on β_0, β_1 and Δ ($p = 0.65$). Thus, selection of the best fitting polynomial, in the presence of information on exposure level, would lead to selection of the linear prevalence log-odds function, and to the corresponding finding of a stronger association between exposure to the test compound and tumour prevalence.

In a simulation study comparing $\hat{X}^2_{PT}$ with linear and quadratic prevalence log-odds functions to X^2_{PT} with a variety of interval selection methods, Dinse (1985) simulated tumour prevalence functions based on Weibull distributed times to tumour. These prevalence functions are clearly not linear in time on the logistic scale. Under the null hypothesis, results of the test based on $\hat{X}^2_{PT}$ with a quadratic prevalence log-odds function agreed more closely with those of tests based on X^2_{PT} using the various interval selection methods than did those of the test based on $\hat{X}^2_{PT}$ using a linear prevalence function. All tests considered in the simulation study tended to reject too often in cases when mortality increased with dose level, the test based on $\hat{X}^2_{PT}$ with a linear prevalence log-odds function rejecting most often. Hence, on the basis of the above examples and of limited simulation results, it would seem prudent to select the degree of the polynomial, $\gamma(t)$, by fitting the model in (5.11) with $\delta = 0$, and to test the significance of successive higher-degree polynomials at a moderate significance level, say 10–20%.

Although the availability of a method for nonlethal tumours which avoids the need for interval selection is desirable, unqualified recommendation of the method of Dinse and Lagakos (1983) must await further investigation of issues related to polynomial selection. Hitchcock (1966) showed that logistic regression tests such as that based on $\hat{X}_{PT}^2$ usually offer only slight gains in efficiency over comparable stratification methods such as tests based on X_{PT}^2, and Gart (1977) verified this finding in a specific application. The simulation study of Dinse (1985) offers further verification. Whatever the slight gain in efficiency of the test based on $\hat{X}_{PT}^2$, it may be offset by the invalidity of the test when the assumed logistic relation over time does not hold (see Cox, 1966). Provided that the tumour prevalence does not change rapidly in any of the selected time intervals, the test based on X_{PT}^2 will be valid regardless of whether a linear relation or some higher polynomial in time holds for tumour prevalence rates. Nevertheless, the logistic regression method of Dinse and Lagakos provides an attractive alternative to interval-based methods, particularly since it provides a statistical framework within which the potential ambiguities introduced by the need to choose intervals or polynomials can be resolved. A recently proposed method based on weighted prevalence estimators (Selwyn *et al.*, 1985) requires neither interval nor polynomial selection, and thus warrants further investigation.

5.6 Analysis of rapidly lethal occult tumours and of observable tumours

In this section we consider statistical methods that use the information on times to tumour, or times to death because of tumour, more precisely. We discuss two kinds of experiments:

(1) studies in which the specific tumour under study is found at necropsy, i.e., is occult, but is assumed to be rapidly lethal, and

(2) studies of easily observable tumours in living animals, such as those in skin-painting experiments, or experiments in which the endpoint is a palpable tumour.

For such studies, we show how to calculate the curves for survival without apparent tumour and give methods for comparing such curves.

In the first kind of study, the tumour when found at death is usually assumed to cause the death, even if the animal died accidentally or was sacrificed because it was moribund. It may be questioned whether animals found with a tumour at scheduled sacrifice should also be assumed to have a lethal tumour. If the tumour is truly rapidly lethal, very few should be found at sacrifice. However, for the terminal sacrifice the fatal/incidental distinction is not essential, since all animals that are killed at this terminal sacrifice were the only ones at risk of dying of a lethal tumour, and considering these tumours as either lethal or incidental will not alter the results. Such questions are discussed more fully in the next section. In any case, the analyses require knowledge of the times of death of all the animals and a categorization of animals into those bearing the tumour of interest and those not bearing the tumour of interest.

Studies in which the tumour is directly observable are simpler to analyse. Here, it is necessary to know the time at which the tumour is first seen in any animal; the times of any subsequent appearance, disappearance, reappearance, appearance of additional tumours, or death are not required for analyses of this type. We do require knowledge of the times of death of all animals that die without ever getting the tumour.

The data are usually recorded in time units of days or weeks and are divided into sets labelled 'uncensored' or 'censored' for animals with and without tumours, respectively. Our analyses require specification of all time points at which tumours are found in any group, denoted, as in Section 5.2, by $t_1, \ldots, t_k, \ldots, t_K$. The number of animals with tumour found at t_k in group i is denoted by y_{ik}. The total number of animals with tumour found over the course of the experiment is then $y_{i.} = \sum_{k=1}^{K} y_{ik}$. Note that $y_{i.}$ corresponds to y_i from Section 5.4, where the time index was suppressed. In general, the number at risk at t_k for group i is m_{ik}. Thus, m_{i1} corresponds to m_i in Section 5.4. Successive values of $m_{i,k+1}$ are found to be $m_{ik} - y_{ik}$ less the number dying in group i in the interval $[t_k, t_{k+1})$. That is, those dying at t_{k+1} are still included in $m_{i,k+1}$. Note that for experiments on lethal tumours, m_{ik} consists of all animals alive at the beginning of t_k, while for observable tumours m_{ik} excludes those living animals that already have a tumour.

We illustrate this notation by considering the data from Section 4.2 on observable tumours induced by painting cigar-smoke condensate in groups D, E and F from Table 4.2. No untreated control group, which would naturally be indexed with 0, is present in this example. We therefore index the three groups 0, 1 and 2, for increasing dose, giving index 0 to the group receiving 59 mg cigar-smoke condensate per week (group F). For numerical convenience, we subtract 59 from the actual doses and code them as 0, 19 and 44. Note that $I = 2$ in this example. Table 5.16 gives a partial listing of the data that have $K = 37$ distinct time points with tumour.

Calculation of the Kaplan–Meier estimator of the survival function without known tumour follows easily from the formulae given in Section 5.3. (It is interesting to note that this procedure was originally used in the same framework for analysing the occurrence of tumours in skin-painting experiments by Miescher et al., 1941.) If y_{ik} is substituted for x_{ik} and m_{ik} for n_{ik}, the formulae for $\hat{S}_i(t)$ and $V\{\hat{S}_i(t)\}$ apply to the

Table 5.16 Summary of ages at detection of observable tumours in data on cigar-smoke condensate

Tumour time point (weeks)		y_{ik}/m_{ik}				Total
		i:	0	1	2	
k	t_k	d_i:	0	19	44	$y_{.k}/m_{.k}$
1	24		1/83	0/81	0/79	1/243
2	31		1/79	0/77	0/73	1/229
3	38		1/76	1/74	1/69	3/219
⋮	⋮		⋮	⋮	⋮	⋮
36	99		0/19	1/9	1/3	2/31
37	100		0/18	0/9	1/2	1/29
Total: $y_{i.}$			21	31	37	$y_{..} = 89$

Fig. 5.2   Kaplan–Meier estimates of tumour-free survival curves for three groups of female
mice (□, 59 mg/week; ○, 78 mg/week; ▲, 103 mg/week) painted with cigar-smoke
condensate

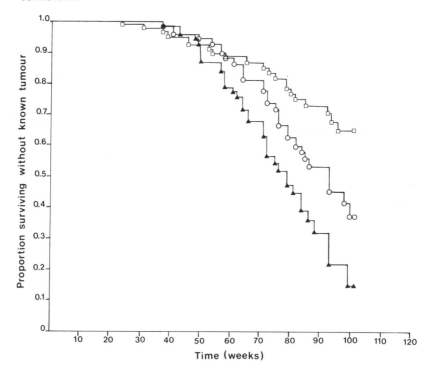

present analysis. The curves are plotted similarly. (See Figure 5.2 for the plot of the
cigar-smoke condensate study.)

The statistical tests for possible differences in these curves of survival without known
tumour are also analogous to the tests of survival curves and of prevalent tumour rates.
Log-rank test statistics may be computed using (5.3), (5.4) and (5.5), where D_i and V_{hi}
are calculated as in (5.1) and (5.2) after substituting y_{ik} for x_{ik} and m_{ik} for n_{ik}. Let X^2_{LH}
denote the test statistic for equality of tumour mortality or incidence rates in the $I + 1$
groups computed using (5.3) (Mantel, 1966; Cox, 1972), X^2_{LT} denote the corresponding
trend test statistic computed using (5.4) (Tarone, 1975), and X^2_{LQ} denote the
corresponding departure from the trend statistic computed using (5.5) (Tarone, 1975).
Furthermore, Z_{LT} may denote the corresponding one-tailed normal deviate for trend.

Alternatively, modified Wilcoxon rank sum tests can be employed. Wilcoxon test
statistics may be computed using (5.8), (5.9) and (5.10), where D_{iW} and V_{hiW} are
calculated as in (5.6) and (5.7) after substituting y_{ik} for x_{ik} and m_{ik} for n_{ik}. As discussed
previously, the modified Wilcoxon test is more sensitive for detecting differences in the
curves early in the experiment. For experiments with heavy intercurrent mortality, the
Peto–Prentice-modified Wilcoxon statistics should be used (Prentice & Marek, 1979).

We turn now to the estimation of the strength of association of tumour effect with
dose. First, consider the assumptions that Cox (1972) made in deriving the log-rank

test. Let $\lambda_i(t)$, the hazard rate function in group i, be the (instantaneous) probability of a tumour in the time interval $(t, t + \Delta t)$. Cox assumed proportional hazard rates, i.e., $\lambda_i(t)/\lambda_0(t) = \rho_i$ for all t and $i = 1, \ldots, I$. That is, the incidence of tumours may vary over time, but the pairwise relative risks among all the groups are constant. The homogeneity chi-square tests whether $\rho_i = 1$ for all i, while the trend test is sensitive to log-linear alternatives of the form $\rho_i = \exp(\beta d_i)$. Various estimators have been suggested for ρ_i. The simplest to use is the ratio of ratios of observed and expected values (Pike, 1972),

$$\hat{r}_i = (O_i/E_{iL})/(O_0/E_{0L}), \qquad i = 1, 2, \ldots, I,$$

where E_{iL} $(i = 0, 1, \ldots, I)$ is the expected number of lethal or observable tumours, calculated as described in general terms in Section 5.2.

Breslow (1975) and Bernstein et al. (1981) found that this estimator is valid for $\frac{1}{2} \le \rho \le 2$, but that it may be biased towards unity for other values. Another rather simply applied estimator is the so-called Mantel–Haenszel estimator, which is essentially a weighted combination of the cross-product ratios:

$$\tilde{r}_i = \left\{ \sum_{k=1}^{K} y_{ik}(m_{0k} - y_{0k})/m_{.k} \right\} \bigg/ \left\{ \sum_{k=1}^{K} y_{0k}(m_{ik} - y_{ik})/m_{.k} \right\}.$$

The results of Bernstein et al. (1981) indicate that this estimator has small bias away from unity for $\frac{1}{3} \le \rho_i \le 3$, but may have large bias otherwise.

Finally, as suggested by Gart (1972), the methods of logistic regression for contingency tables can be applied to either the $2 \times 2 \times K$ or $2 \times (I+1) \times K$ table. From the former, an estimator of ρ_i for each $i = 1, 2, \ldots, I$ can be found from the maximum likelihood estimator, $\hat{\rho}_i$ of the odds ratio (see, e.g., Thomas, 1975). Alternatively, we can fit a stratified logistic model to the $2 \times (I+1) \times K$ table and obtain the maximum likelihood estimator of the common slope, $\tilde{\beta}$ (see, e.g., Thomas & Gart, 1983). The estimators of ρ_i based on the linear logistic model are then $\tilde{\rho}_i = \exp(\tilde{\beta} d_i)$. Either of these estimators tends to be biased away from unity with small sample sizes.

Let us return now to the example of painting with cigar-smoke condensate. We find the one-tailed test for trend is $Z_{LT} = 4.49$, $p = 0.000004$. The comparable test statistic using the Wilcoxon form of the test yields a normal deviate Z_{LT} of 3.46 with $p = 0.00027$. The Wilcoxon form is less significant, as it gives greater weight to the comparisons early in the experiment, where the curve of survival without tumour of the control group is actually lower than that of the dosed groups.

The full set of chi-square analyses by both tests is given in Table 5.17. Chi-squares for heterogeneity are highly significant by both methods. Almost all of this variation is accounted for by the linear trend chi-squares. Thus, neither method finds any evidence of departure from linearity.

Consider now the question of estimating the strength of association between dose and tumour effect, illustrated for this example in Table 5.18. For this example, $\tilde{\beta} = 0.0271 \pm 0.0062$. The good agreement between $\tilde{\rho}_i$, which assumes linearity on the logistic scale, and the other estimates, which do not, reflects the low chi-squares for departure from linearity.

Table 5.17 Summary of analysis of observable tumours from data on cigar-smoke condensate

Source of variation	Degrees of freedom	Log-rank test		Wilcoxon test	
		Chi-square	p	Chi-square	p
Linear trend	1	20.15	0.0000	11.96	0.0005
Departure from linearity	1	0.01	0.9187	0.06	0.8124
Total (heterogeneity)	2	20.16	0.0000	12.02	0.0025

Table 5.18 Estimates of odds ratios from data on cigar-smoke condensate

	Dose d_i		
	0	19	44
$O_i = y_i$.	21	31	37
E_{iL}	37.92	29.74	21.35
O_i/E_{iL}	0.554	1.042	1.733
$\hat{r}_i$	1.00	1.88	3.13
$\tilde{r}_i$	1.00	1.88	3.12
$\hat{\rho}_i$	1.00	1.91	3.34
$\tilde{\rho}_i = \exp(\tilde{\beta}d_i)$	1.00	1.67	3.29

Comparison of life-table analyses and analyses of crude proportions

It is not uncommon for analyses of crude proportions to yield substantially the same interpretation as that reached by the more elaborate analyses using the life-table techniques we have just described. If there is more intercurrent mortality among the high-dose groups (see Chapter 2, Table 2.2), the more sophisticated analyses will probably yield a more significant positive or direct association with dose than will the crude analysis. Under these circumstances, existing relationships not found by the crude analysis may become apparent in the life-table analysis.

If the intercurrent mortality is minimal or roughly equal in the various groups, the crude proportion and life-table analyses will often be quite similar. Cuzick (1982) confirmed this impression theoretically. He found that crude proportion analysis is over 95% efficient relative to Cox's life-table analysis when less than 50% of the animals have tumours.

It is instructive to compare the results of the crude proportion analysis for the example of cigar-smoke condensate. Note that the first tumour occurred at 24 weeks and the last at 100 weeks. The summary of the data on early and intercurrent mortality is given in Table 5.19. We see that, even after adjusting for early mortality, there remain substantial differences in the percentages of intercurrent mortality among the groups of those at risk at 24 weeks and not getting a tumour subsequently. These range from 74% to 98%.

Table 5.19 Summary of mortality from data on cigar-smoke condensate

	Dose d_i			Total
	0	19	44	
n_i	100	100	100	$n_. = 300$
Death before $t = 24$	17	19	21	57
m_{i1}	83	81	79	$m_{.1} = 243$
Interim deaths (24–100 weeks)	46	46	41	133
y_i	21	31	37	
$m_{i1} - y_i$	62	50	42	
Interim deaths (% of $m_{i1} - y_i$)	74%	92%	98%	

For both the life table and the crude proportions tests, we use the identical total observed numbers of animals with tumours, but the expected values are calculated differently (see Section 5.4). Table 5.20 presents the results.

Table 5.20 Observed and expected values calculated by two methods (life-table and crude proportions) from data on cigar-smoke condensate.

	Dose d_i			Total
	0	19	44	
$O_i = y_i$	21	31	37	$y_{..} = 89$
E_{iL}	37.42	29.74	21.35	89
$O_i - E_{iL}$	−16.92	1.26	15.65	0
$E_i = (m_{i1}y_{..})/m_{.1}$	30.40	29.67	28.93	89
$O_i - E_i$	−9.40	1.33	8.07	0

Because of the high intercurrent mortality in the high-dose group, its expectation is much lower in the life-table analyses than it is in the crude proportion analysis, which does not take the differential mortality into account. The inverse relation holds for the low-dose group, which has lower intercurrent mortality. Thus, deviations of the expected from the observed values are larger in the life-table analysis. The results of the crude proportion analysis are shown in Table 5.21.

Table 5.21 Summary of crude proportion analysis of data on cigar-smoke condensate

Source of variation	Degrees of freedom	Crude proportion tests	
		Chi-square	p
Linear trend	1	7.88	0.0050
Departure from linearity	1	0.31	0.5771
Total (heterogeneity)	2	8.19	0.0166

Table 5.22 Estimates of odds ratio from crude proportion
analysis of data on cigar-smoke condensate

Dose	d_i	0	19	44
Odds ratio	$\hat{R}_i$	1.00	1.83	2.60
	$\bar{R}_i$	1.00	1.49	2.52

The trend and homogeneity tests are still significant but less so than by either of the life-table analyses. Recalling the discussion of Section 5.4, the associated measures of association are shown in Table 5.22. For this analysis, we found that $\bar{\beta} = 0.0210 \pm 0.0075$. These estimates are somewhat lower, particularly at the high dose, than the estimates found from the life-table analysis. Thus, although both analyses imply similar qualitative interpretations, the life-table analyses yield more highly significant tests with larger measures of association. This is a quite common result and represents an example of outcome type A of Table 2.2.

Alternative life-table and exact tests

When the experiment is small, the question arises whether the life-table tests are valid. Simulations performed by Tarone (1975) and Latta (1981) indicate that the fit of the test statistics to the appropriate chi-square distributions is quite good in large sample sizes. However, when the numbers are small, these tests may reject the null hypothesis too often. For small expected values, say, the minimum of $E_{iL} < 5$, a conservative version of the life-table chi-square tests has been suggested (Peto & Pike, 1973). However, Gart (1975) and Haybittle and Freedman (1979) point out circumstances in which application of the conservative test must be used with caution. If only one of the groups, say, i, has animals at risk beyond time t', then this should be made the final time, t_K, for the analyses based on the conservative test. If tumours occur for $t > t'$ in the ith group, then their time points are omitted from calculation of the observed and expected. Thus, $O_i < y_i$. in such instances.

For very small numbers, exact versions of these tests can, in principle, be done. However, this involves an additional assumption requiring that the hazard rate for deaths without tumour in each group be proportionally related to its hazard rate for tumour incidence. Such tests are described by Tarone (1975) and Cox (1959).

A test for acceleration (as defined in Section 2.2) has been developed by Breslow *et al.* (1984). Their test statistic is of the form given in equation (5.6), with w_k taken to be the estimated cumulative hazard function (using the pooled data from control and exposed groups) at t_k, the kth ordered time of death due to tumour. Because acceleration is unlikely in experiments with inbred strains, this test should be used in conjunction with a test such as the log-rank test which has power against more general alternatives. Accordingly, appropriate adjustment for multiple comparisons is required in those cases in which the acceleration test is employed (Breslow *et al.*, 1984).

Combination of tests

The approximate test for trend can easily be combined over experiments, as with the analysis of crude proportions. One simply adds the numerators of the Z_{TL} and divides by the square root of the sums of squares of the individual denominators. The resulting statistic is an approximate normal deviate. Combination of the homogeneity and departure chi-squares usually requires a computer program (see, e.g., Thomas & Gart, 1983).

Issues of multiple comparisons

When the statistical analysis is done by comparing in turn each dose group, d_i, to the control, d_0, the question of multiple testing can be handled by the Bonferroni inequality, as it was for the analysis of crude proportions (see Section 5.4). In addition, because we have suggested two life-table adjusted tests for trend, another question of multiple testing arises. It is invalid to compute routinely both tests and report only the one that yields the higher significance. Tarone (1981) has given a method for adjusting the *p*-value for the more extreme of these two tests. The adjustment method is derived explicitly for the Breslow-modified Wilcoxon; however, the method applies also to the Peto–Prentice-modified Wilcoxon, after substituting the Peto–Prentice weights for the Breslow weights throughout.

5.7 Analysis of occult tumour data when contexts of observation are known

Although the analysis described in Section 5.5 is valid if all tumours of a particular type are nonlethal and the analysis described in Section 5.6 is valid if all tumours of a particular type are rapidly lethal, the relationship between the presence of a tumour and death of host animals often lies between these two extremes. Some tumours of a particular type may be, for all practical purposes, nonlethal, while other tumours of the same type may contribute to the death of their host animals. As noted in Chapter 2, differences among exposure groups with respect to intercurrent mortality can cause serious bias in tests for equality of tumour rates. For data on occult tumours, analysis assuming that all tumours are lethal when, in fact, some are nonlethal, and analysis assuming that all tumours are nonlethal when, in fact, some are lethal, can lead to incorrect inferences if intercurrent mortality rates differ among exposure groups (Peto *et al.*, 1980; Lagakos, 1982). In the common situation in which intercurrent mortality rates increase with increasing dose level, the tumorigenic effect will be overstated in the first case (i.e., assuming all tumours are lethal when some are nonlethal) and understated in the second case (i.e., assuming all tumours are nonlethal when some are lethal).

In order to avoid biases due to differences in intercurrent mortality and at the same time to make some use of data on time of death, Peto (1974) and Peto *et al.* (1980) recommended that pathologists assign a context of observation to each observed tumour (Section 2.10). A tumour that either directly or indirectly kills its host is said to be observed in a fatal context. A tumour that is observed at necropsy of an animal that died of some cause unrelated to the tumour is said to be observed in an incidental context. Although it may be difficult to determine the contexts of observation for some tumours, it is assumed in this section that all tumours of a particular type have been classified as either fatal or incidental.

The analysis of data on occult tumours using contexts of observation is based on the methods given in Sections 5.5 and 5.6. The analysis of incidental tumours is a straightforward modification of the analysis of nonlethal tumours presented in Section 5.5. A modification is necessary, because those animals killed by the tumour in question (i.e., animals for which the tumour is observed in a fatal context) should not enter into the analysis of incidental tumours. As in the analysis of Section 5.5, the length of the experiment is subdivided into distinct time intervals. Within each time interval, the data on incidental tumours may be summarized in a table such as Table 5.3, with N_{ij} corresponding to the number of animals in group i dying during interval j from causes unrelated to the presence of the tumour in question, and Y_{ij} corresponding to the number of these animals in which the tumour was observed in the incidental context, for $i = 0, 1, 2, \ldots, I$ and $j = 1, 2, \ldots, J$. All tumours found in animals killed in planned sacrifices are classified as incidental. Using the methods of Section 5.3, we form a vector $\mathbf{D}_P$ of differences of expected from observed values for the data on incidental tumours and compute the corresponding covariance matrix $\mathbf{V}_P$.

Analysis of tumours observed in the fatal context is based on the methods in Section 5.6. At each time t_k at which the tumour in question is observed in the fatal context, a contingency table like Table 5.16 can be formed, where y_{ik} corresponds to the number of animals in group i for which the tumour was observed in the fatal context at t_k, and m_{ik} corresponds to the number of animals in group i surviving (and thus still at risk of being killed by a tumour) to time t_k. Note that, in the analysis of fatal tumours, animals in which the tumour is observed in the incidental context are treated exactly as all other animals not killed by the tumour. As in Section 5.6, a vector $\mathbf{D}_L$ of differences of expected from observed values is formed using the fatal tumour data, and the corresponding covariance matrix $\mathbf{V}_L$ is computed.

The analysis of data on occult tumours using contexts of observation is based on the vector $\mathbf{D}_C = \mathbf{D}_P + \mathbf{D}_L$, with covariance matrix $\mathbf{V}_C = \mathbf{V}_P + \mathbf{V}_L$. Test statistics for heterogeneity X^2_{CH}, trend X^2_{CT}, and departure from trend (X^2_{CQ}) may be calculated as in (5.3), (5.4) and (5.5), respectively, with $\mathbf{D}_C$ substituted for $\mathbf{D}$ and $\mathbf{V}_C$ substituted for $\mathbf{V}$.

As noted earlier, it may be difficult to determine the contexts of observation of some tumours. Accordingly, Peto *et al.* (1980) suggest that tumours be classified on an ordinal scale, namely, 1 if a tumour is definitely incidental, 2 if a tumour is probably incidental, 3 if a tumour is probably fatal and 4 if a tumour is definitely fatal. Usually, the above analysis would be performed with tumours classified as 1 or 2 taken to be incidental and tumours classified as 3 or 4 taken to be fatal. With the ordinal classification, however, the analysis can be repeated using different cut points (e.g., classifications 1, 2 and 3 taken as incidental and classification 4 taken as fatal) to determine if inferences are influenced by possible misclassification of tumour context. For further discussions of the problems associated with the assignment of causes of death, see Lagakos (1982) and Kodell *et al.* (1982b).

Consider the data on pituitary tumours given in Table 4.3 for 16 groups of male Colworth rats exposed to increasing dose levels of N-nitrosodimethylamine. As noted by Peto *et al.* (1980), treatment-induced, fatal liver tumours caused a marked, dose-related increase in mortality. Pituitary tumours were observed in both fatal and incidental contexts and were classified on the ordinal scale described in the preceding paragraph. The first part of Table 5.23 gives the observed and expected numbers of animals with tumours found in a fatal context; these are tumours recorded as fatal, or probably fatal, with the codes 4 or 3 in Table 4.3. The observed and expected numbers were derived using the methods of Section 5.6, and the vector of their difference is $\mathbf{D}_L$, the corresponding covariance matrix $\mathbf{V}_L$ not being displayed.

The code -3 in Table 4.3 was used for animals that were totally cannibalized or autolysed; their cause of death was not ascertainable. These animals were considered in the analysis of fatal tumours as if they had died on day 1 without a tumour and were excluded from the incidental tumour analysis.

The code -2 in Table 4.3 was used for animals whose head was cannibalized or autolysed so that the presence or absence of a pituitary tumour was not ascertainable but death was known not to be caused by a pituitary tumour. These animals were considered with their respective time to death as having no fatal pituitary tumour in the analysis of fatal tumours but were excluded from the analysis of incidental tumours.

For the prevalence analysis, the adaptive method of determining subintervals of the time axis, outlined in Section 5.5, was used and resulted in the following ten intervals:

$$[87, 591], [596, 680], [681, 796], [797, 891], [892, 905], [908, 964],$$
$$[965, 1028], [1029, 1030], [1033, 1071], [1073, 1234].$$

The 17 animals with codes -2 and -3 were, as explained above, excluded from the prevalence analysis. Thus, the number of animals considered in each group for the analysis of incidental tumours is not always equal to the number of animals considered in the analysis of fatal tumours less the number of fatal tumours observed.

The second part of Table 5.23 gives the calculated observed and expected numbers for each group in the prevalence analysis. Their difference is the vector $\mathbf{D}_P$, the corresponding covariance matrix $\mathbf{V}_P$ not being displayed.

In the third part of Table 5.23, the numbers of tumours observed and expected in either context are summed for each group. The difference is the vector $\mathbf{D}_C$, the corresponding covariance matrix being $\mathbf{V}_C = \mathbf{V}_L + \mathbf{V}_P$.

The chi-square statistic for heterogeneity calculated according to (5.3) for the combined situation is $X^2_{CH} = 12.11$. Because all animals in the highest dose group died before any pituitary tumour was observed in the experiment, the observed and expected numbers are zero. The expected number of tumours in group 15 is virtually zero ($E_{15} = 0.02$), and thus group 15 was also deleted prior to the computation of test statistics. The rank of the matrix $\mathbf{V}_C$ is therefore 13 rather than 15. Comparison of the computed X^2_{CH} to the percentiles of a chi-square distribution with 13 degrees of freedom gives a p-value of 0.52.

Using the scores $0, 1, \ldots, 12, 13$ as dose levels, corresponding roughly to a

Table 5.23 Pituitary tumours in male Colworth rats: observed and expected numbers of tumours by context of observation and experimental group

Group	Tumours observed in a fatal context			Tumours observed in an incidental context			Combined	
	No. of animals	Observed events	Expected events	No. of animals	Observed events	Expected events	Observed events	Expected events
1	192	26	32.24	159	24	24.12	50	56.36
2	48	10	9.44	38	6	6.44	16	15.88
3	48	8	8.11	40	6	5.96	14	14.07
4	48	5	8.45	41	9	6.63	14	15.08
5	48	8	7.77	39	3	5.46	11	13.23
6	48	10	9.48	38	8	6.82	18	16.30
7	48	9	7.14	39	3	5.14	12	12.28
8	48	11	5.92	36	6	4.76	17	10.68
9	48	5	7.09	43	7	6.22	12	13.31
10	48	6	3.43	40	5	3.41	11	6.83
11	48	5	2.70	43	1	2.71	6	5.41
12	48	1	1.85	47	1	2.33	2	4.18
13	48	0	0.23	48	2	0.59	2	0.82
14	48	0	0.14	48	0	0.41	0	0.55
15	48	0	0.02	47	0	0.00	0	0.02
16	48	0	0.00	45	0	0.00	0	0.00

logarithmic transformation of the actual dose levels (see Section 4.3), gives, according to (5.4), a trend statistic $X^2_{CT} = 1.66$ with a two-sided p-value of 0.20.

Thus, it appears that *N*-nitrosodimethylamine does not induce pituitary tumours in male Colworth rats.

This data set is of particular interest since, if one did not use the information on the context of observation but considered all the pituitary tumours to be found either in a fatal context or in an incidental context, different conclusions would be derived.

Considering all pituitary tumours as fatal would give, with the life-table methods of Section 5.6, a chi-square statistic for heterogeneity $X^2_{LH} = 28.34$ with 13 degrees of freedom ($p = 0.008$). The one-degree-of-freedom chi-square statistic for trend would be $X^2_{LT} = 7.38$ ($p = 0.007$), indicating a positive trend in the occurrence of pituitary tumours with increasing dose. Considering all tumours as incidental would give, using the prevalence methods of Section 5.5, a heterogeneity statistic $X^2_{PH} = 18.04$ with 13 degrees of freedom ($p = 0.156$). The one-degree-of-freedom chi-square statistic for trend would be $X^2_{PT} = 2.85$ ($p = 0.091$), suggesting a negative trend in occurrence of pituitary tumours with increasing dose.

As noted in Chapter 2, the combination of two analyses, one based on tumour death rates and the other based ostensibly on tumour prevalence rates, may seem somewhat contrived. In fact, it is difficult to justify this analysis rigorously. The tests based on X^2_{CH} and X^2_{CT} can be shown to test the null hypothesis of interest, that is, that the underlying tumour onset rates are equal in all exposure groups, only under rigid assumptions (Lagakos, 1982; McKnight & Crowley, 1984). Nevertheless, this analysis is a useful attempt to solve the difficult problem of using data on age at death in evaluating data on occult tumours. Because the underlying tumour onset rates are not identifiable (Lagakos, 1982; McKnight & Crowley, 1984), any test for equality of onset rates will be only approximate, and efforts to improve upon the analyses in this section are likely to require changes in experimental design. For example, McKnight and Crowley (1984) have shown that tumour onset rates are approximately identifiable in experiments with frequent planned sacrifices. Thus, it is likely that better tests can be developed, but only at the cost of additional animals. Methods of testing for differences in tumour incidence rates using data from planned sacrifices are now available (McKnight & Crowley, 1984; Dewanji & Kalbfleisch, 1986), and research in this area is progressing rapidly.

LIST OF ESSENTIAL SYMBOLS – CHAPTER 5 (in order of appearance)

$I+1$	number of experimental groups ($i = 0, 1, \ldots, I$)
d_i	dose level ($d_0 = 0, d_1 < \cdots < d_I$)
t_k	time of observation of an event (death, occurrence of tumour) ($k = 1, \ldots, K$)
x_{ik}	number of events in group i at time t_k
n_{ik}	number of animals at risk (to experience event) in group i at time t_k
A_{ik}	proportion of animals at risk in group i compared to total of all groups at time t_k
E_{ik}	expected number of events in group i at time t_k
O_i	total number of events in group i
E_i	total number of expected events in group i
D_i	difference between O_i and E_i
$\mathbf{D}$	vector of D_i's
$\mathbf{V}$	covariance matrix of vector $\mathbf{D}$
V_{hi}	element of $\mathbf{V}$ ($h, i = 0, 1, \ldots, I$)
X_H^2	test statistic for heterogeneity (chi-squared distribution with I degrees of freedom)
X_T^2	two-sided test statistic for linear trend (chi-squared distribution with 1 degree of freedom)
X_Q^2	test statistic for departure from linear trend (chi-squared distribution with $I-1$ degrees of freedom)

(N.B.: A subscript W to the above quantities D_i, $\mathbf{D}$, $\mathbf{V}$, V_{hi}, X_H^2, X_T^2 and X_Q^2 indicates their derivation using non-negative weights w_k ($k = 1, \ldots, K$))

$\hat{S}_i(t)$	Kaplan–Meier estimate of survival function in group i
$V\{\hat{S}_i(t)\}$	variance of $\hat{S}_i(t)$
y_i	number of animals with tumour in group i ($i = 0, 1, \ldots, I$)
m_i	number of animals at risk in group i ($i = 0, 1, \ldots, I$)
$\hat{R}_i$	ratio of odds of a tumour in group i to corresponding odds in control group
Z	one-tailed test statistic for difference of two proportions (normal distribution); subscript c when used with continuity correction

Z_T one-tailed test statistic for monotone trend (normal distribution); subscript c when used with continuity correction

$\tilde{R}_i$ odds ratio estimate from regression model

T duration of experiment

$\mathcal{I}_j$ subinterval of experimental time spanning from 0 to T ($j = 1, \ldots, J$)

Y_{ij} number of animals with tumour among those that died or were killed in group i in interval j

N_{ij} number of animals that died or were killed in group i in interval j

(N.B.: Subscript P, L and C given to quantities X_H^2, X_T^2 and X_Q^2 when derived in prevalence analysis (Section 5.5), life-table analysis (Section 5.6) or analysis using context of observation (Section 5.7))

$\lambda_i(t)$ hazard rate function in group i

ρ_i ratio of hazard rates between group i and control group: $\lambda_i(t)/\lambda_0(t)$

$\hat{r}_i$ estimate of ρ_i by ratio of observed and expected values ($i = 1, \ldots, I$)

$\tilde{r}_i$ estimate of ρ_i by weighted odds ratios ($i = 1, \ldots, I$)

6. MODEL FITTING

CHAPTER 6

MODEL FITTING

6.1 Introduction

As most animal carcinogenesis experiments aim at determining whether or not a particular treatment increases the risk of cancer at one or more sites, the statistical methods described so far lean heavily towards techniques for hypothesis testing. However, many long-term animal experiments are analysed to provide an appropriate condensation of the information contained in the data and not merely to answer a yes–no question.

Many statistical models fitted to experimental data have fruitfully influenced the thinking on chemical carcinogenesis. The multistage model proposed by Armitage and Doll (1961) has been able to account for various experimental as well as epidemiological observations. Age-specific incidence of tumours induced by continuous exposure to chemical carcinogens has been described successfully by such models (Lee & O'Neill, 1971; Berry & Wagner, 1969), which also predicted increasing incidence with age as being a consequence of prolonged time since first exposure. This prediction was confirmed experimentally (Peto et al., 1975).

On the epidemiological side, a marked regularity of age-incidence curves for most spontaneous human tumours of epithelial origin has been noted (Cook et al., 1969) and detailed dose-response curves, as observed for the dependence of lung cancer risk on daily cigarette consumption, show a high degree of consistency with the multistage theory (Doll, 1971; Peto, 1977; Doll & Peto, 1978). Epidemiological data on the joint effect of two exposures are also interpretable in terms of the multistage model (Wahrendorf, 1984). Day and Brown (1980) have explored the multistage model concerning changes in risk after cessation of exposure. They were able to show both for experimental and epidemiological data two different types of behaviour under the multistage model.

An interesting empirical observation for chemical carcinogenesis data was made by Druckrey et al. (1967). They noted that if d is the daily dose rate and t the median time to tumour induction (death with tumour), then the relationship, $d \cdot t^n = $ constant, holds for many carcinogens, especially nitroso compounds. This formula is predicted by the Weibull model (Carlborg, 1981) and has received wide attention in the discussion of thresholds in chemical carcinogenesis (Chand & Hoel, 1974; Port et al., 1976).

The fitting of statistical models has to compromise between specificity and identifiability. On the one hand, one would imagine that inserting all relevant knowledge about the carcinogenic process into a mathematical model would result in many parameters which could not all be identified by the limited amount of information provided from a long-term animal experiment. On the other hand, models which relate the probability of tumour development throughout life to the dose administered appear very simple.

The degree of specificity which a model might be allowed to have depends on the details of the experimental design, including whether and when animals are inspected for diagnostic purposes and whether and when animals are killed – either by scheduled sacrifice or by normally occurring deaths. The experimental design also specifies the schedule of the dose application. Of special interest in this respect are chronic exposures stopped at a certain time or the application of fractionated doses which can differ in respect to total dose, number of fractions and time between single doses.

One essential problem with the fitting of statistical models should be stated clearly. These models are usually fitted to data sets from single experiments done in one strain, in one sex, by one route of application and by considering tumours at one site. Consequently, the scope of generalization of one such fit is limited, particularly in the absence of consistency in studies done under different conditions.

In the following sections, we will give an overview of some statistical models which have been proposed for the analysis of long-term animal experiments. This will include simple dose-response models, time-to-tumour models and models based on different states in the course of the tumour development.

6.2 Dose-response models

As noted in Chapter 3, the probability of tumour occurrence depends on both the dose and the period of exposure. In many cases, however, the dose-response relationship at a fixed point in time may be of interest. In the absence of decreased survival at high doses, for example, the proportion of animals developing tumours during the course of a long-term study generally increases with dose.

The shape of such dose-response curves can vary widely depending on the agent used (Fig. 6.1). While the dose-response curve for liver tumours induced in mice as a result of exposure to 2-acetylaminofluorine (2-AAF) for 24 months is nearly linear (Littlefield et al., 1980a), other curves may be distinctly nonlinear. The dose-response curve for liver tumours induced as a result of exposure to gaseous vinyl chloride (six hours per day for two years) increases somewhat linearly at low doses and then tends to level off at higher doses (Maltoni, 1975). This plateau effect is thought to be due to saturation of the metabolic activation mechanism for vinyl chloride in the liver (Gehring et al., 1978). Conversely, the dose-response curve for squamous-cell carcinomas of the nasal passage induced as a result of exposure to gaseous formaldehyde (also six hours per day for two years) shows a marked increase in response above 5.6 ppm (Swenberg et al., 1983), possibly due to saturation of the mucociliary clearance mechanism. Finally, the dose-response curve for liver tumours resulting from ingestion of aflatoxin in the diet (Wogan et al., 1974) increases at low doses but levels off at high doses as a response rate approaching 100% is reached.

These examples clearly illustrate that the dose-response curves for different chemical

Fig. 6.1 Examples of dose-response relationships for vinyl chloride (from Maltoni, 1975), aflatoxin (from Wogan *et al.*, 1974), 2-acetylaminofluorene (from Littlefield *et al.*, 1980a) and formaldehyde (from Swenberg *et al.*, 1983)

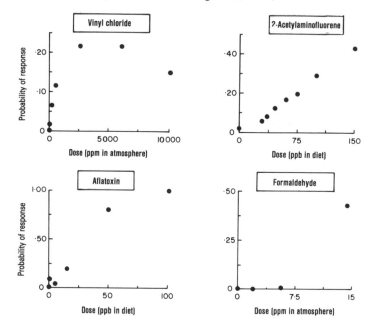

agents can be quite dissimilar. This kind of variation in shape is also encountered in radiation carcinogenesis (Ullrich *et al.*, 1976; Ullrich & Storer, 1979a,b; Ullrich, 1980). In the remainder of this section, we consider the problem of modelling the dose-response relationship for a given compound in order to obtain a more quantitative description of the data. We shall consider also the use of such models in estimating the response rate at doses not included in the experimental protocol.

Some mathematical models

The relationship between the crude proportion of animals developing tumours during the course of a bioassay and the level of exposure may be described by means of a statistical model relating the probability of tumour induction $P(d)$ and the dose d. Statistical or tolerance distribution models are based on the concept that each animal has its own tolerance to the test compound and will develop a lesion only if that tolerance is exceeded. The tolerances are presumed to vary within the population according to some tolerance distribution, $G(t)$, so that the probability that an animal selected at random will respond to dose d is given by

$$P(d) = \Pr\{\text{tolerance} \le d\} = G(d). \tag{6.1}$$

A general class of tolerance distribution models is defined by

$$G(t) = F(\alpha + \beta \log t), \tag{6.2}$$

where F denotes some suitable cumulative distribution function and α and $\beta > 0$ are parameters (Chand & Hoel, 1974). Three commonly encountered models in this class are the probit, logit and extreme value, defined by

$$F(x) = (2\pi)^{-1/2} \int_{-\infty}^{x} \exp(-u^2/2) \, du, \tag{6.3}$$

$$F(x) = [1 + \exp(-x)]^{-1}, \tag{6.4}$$

and

$$F(x) = 1 - \exp\{-\exp(x)\} \tag{6.5}$$

respectively. Since under the extreme value model $G(t) = 1 - \exp(-at^b)$, where $a = \exp(\alpha)$ and $b = \beta$, this model is sometimes called the Weibull model (see Section 6.3).

Stochastic or mechanistic models are based on the concept that a toxic response is the result of the random occurrence of one or more fundamental biological events. Under the multi-hit model, for example, a response is assumed to be induced once the target tissue has been 'hit' by $k \geq 1$ biologically effective units of dose within a specified time period. Assuming that the number of hits during this period follows a Poisson process, the probability of a response is given by

$$P(d) = \Pr\{\text{at least } k \text{ hits}\} = 1 - \sum_{j=0}^{k-1} \exp(-\lambda d) \frac{(\lambda d)^j}{j!}, \tag{6.6}$$

where $\lambda d > 0$ denotes the expected number of hits during this period (Rai & Van Ryzin, 1981). When $k = 1$, the multi-hit model reduces to the one-hit model given by

$$P(d) = 1 - e^{-\lambda d}. \tag{6.7}$$

Incorporation of background response

All of the above models imply that the background response rate $P(0)$ is zero. In many cases, however, the response of interest will also occur spontaneously in control animals (Tarone *et al.*, 1981). Spontaneously occurring lesions may be assumed to arise as a result of a variety of biological mechanisms. Two commonly encountered assumptions in this regard are independence and additivity (Hoel, 1980). In the first case, spontaneous and induced lesions are presumed to occur independently of each other so that the probability of either a spontaneous or a treatment-induced response occurring is given by

$$P^*(d) = \pi_0 + (1 - \pi_0)P(d), \tag{6.8}$$

where $0 < \pi_0 < 1$ denotes the background rate of response (Abbott, 1925). Under additivity, spontaneously occurring effects are considered to be due to an effective background dose $\delta > 0$, with

$$P^*(d) = P(d + \delta). \tag{6.9}$$

Note that, with the one-hit model, the independence and additivity assumptions are indistinguishable. A combination of both independent and additive background may be represented by the model

$$P^*(d) = \pi_0 + (1 - \pi_0)P(d + \delta). \tag{6.10}$$

Other models

A simple class of tolerance distribution models in which background response arises in neither an independent nor additive fashion is defined by

$$P(d) = F(\alpha + \beta d), \tag{6.11}$$

with $\beta > 0$ as in (6.2). When F follows the logistic distribution in (6.4), Cox (1970) has shown that the uniformly most powerful, unbiased test for positive slope β in the proportion of animals responding with increasing dose is the Cochran–Armitage test discussed in Chapter 5. Subsequently, Tarone and Gart (1980) demonstrated the robustness of this test by showing that it is the locally most powerful for any monotone increasing distribution F. However, because the model given by (6.11) involves only two parameters, it is less flexible than those given by (6.8), and may not always provide an adequate description of the observed dose-response curve.

Armitage–Doll multistage model

Perhaps the most widely applied model in the case of carcinogenesis is the Armitage–Doll multistage model (Armitage, 1982). In this case, it is assumed that a cell line progresses through k distinct stages prior to becoming cancerous and that the rate of occurrence of the ith change is of the form $\lambda_i = \alpha_i + \beta_i d$, where $\alpha_i > 0$ and $\beta_i \geq 0$ for $i = 1, \ldots, k$. The parameter α_i represents the spontaneous rate of occurrence of the ith change in the absence of any exposure, and the rate is supposed to be linearly dependent on dose through $\beta_i d$. The probability of a response within a given time period is then approximately

$$P^*(d) = 1 - \exp\left\{-c \prod_{i=1}^{k} (\alpha_i + \beta_i d)\right\}, \tag{6.12}$$

where $c > 0$ (Crump *et al.*, 1976). Noting that the exponent in (6.12) is a polynomial in dose, this model can also be viewed as

$$P^*(d) = 1 - \exp\left\{-\sum_{i=0}^{k} b_i d^i\right\}, \tag{6.13}$$

where the b_i are subject to certain nonlinear constraints (Krewski & Van Ryzin, 1981). For simplicity, however, the linear constraints $b_i \geq 0$ are often employed in practice, providing a more general model than given by (6.12) (Crump *et al.*, 1977).

Pharmacokinetic model

In many cases, a chemical will require some form of metabolic activation before it may exert its toxic effects (Cornfield, 1977). Rai and Van Ryzin (1983), for example,

Fig. 6.2 A simple pharmacokinetic model for metabolic fate of a compound

consider the simple compartmental model shown in Figure 6.2. Here, the administered dose $D_1(t)$ at time t undergoes a transformation T_1 to the activated form $D_2(t)$ and may then be eliminated *via* a second transformation T_2. Each transformation T_i is assumed to follow saturable Michaelis–Menten kinetics (Karlson, 1965, p. 80), with

$$\text{Rate }(T_i) = \frac{b_i D_i(t)}{c_i + D_i(t)}, \tag{6.14}$$

where $b_i, c_i > 0 \; (i = 1, 2)$. Assuming that the dose is administered at a constant rate k, the system satisfies the nonlinear differential equations

$$\frac{dD_1(t)}{dt} = k - \frac{b_1 D_1(t)}{c_1 + D_1(t)} \tag{6.15}$$

and

$$\frac{dD_2(t)}{dt} = \frac{b_1 D_1(t)}{c_1 + D_1(t)} - \frac{b_2 D_2(t)}{c_2 + D_2(t)}. \tag{6.16}$$

Under the steady state conditions $dD_i(t)/dt = 0$, it follows from (6.16) that

$$D_2^* = \frac{a_1 d}{1 + a_2 d}, \tag{6.17}$$

where $d = D_1(t)$ is constant, $a_1 = b_1 c_2 / b_2 c_1 > 0$ and $a_2 = (b_2 - b_1)/b_2 c_1 > -1/M$, with M being the highest dose $D_1(t)$ such that the rate of T_1 does not exceed the rate of T_2. If both transformations follow linear kinetics, then $a_2 = 0$ in (6.17) and the effective dose D_2^* is directly proportional to the administered dose d.

The probability of a response is assumed to depend only on the steady-state level of the effective dose D_2^* in (6.17), say

$$P(d) = F[D_2^*(d)]. \tag{6.18}$$

Taking F to be of the one-hit form with an additive constant

$$F(x) = 1 - \exp\{-(\alpha + \beta x)\} \tag{6.19}$$

yields the dose-response model

$$P(d) = 1 - \exp\left\{-\left[\theta_1 + \theta_2\left(\frac{d}{1 + \theta_3 d}\right)\right]\right\}, \tag{6.20}$$

where $\theta_1 = \alpha > 0$, $\theta_2 = \beta a_1 > 0$ and $\theta_3 = a_2 > -1/M$. This model can, depending on the rate coefficients governing T_1 and T_2, describe both downward bending curves, such as that noted for vinyl chloride (saturable activation), as well as 'hockey-stick' shaped curves, such as that for formaldehyde (saturable elimination). Thus, even though the dose-response curve follows a simple one-hit model in terms of the effective dose D_2^*, a variety of curves may still arise as a result of the saturability of the activation and elimination steps.

More generally, Rai and Van Ryzin (1983) also consider F to be of the form

$$F(x) = 1 - \exp\{-(\alpha + \beta x^\gamma)\},\tag{6.21}$$

with $\gamma > 0$. In this case, the overall dose-response model in (6.20) becomes

$$P(d) = 1 - \exp\left\{-\left[\theta_1 + \theta_2\left(\frac{d}{1 + \theta_3 d}\right)^{\theta_4}\right]\right\},\tag{6.22}$$

where $\theta_4 = \gamma$. The use of the additional parameter θ_4 allows for additional curvature in the model $P(d)$ which cannot be accommodated by the pharmacokinetic parameters θ_2 and θ_3.

Gehring and Blau (1977) considered the somewhat more complex model shown in Figure 6.3. Once taken up by the body, a chemical C may be either eliminated immediately or activated to form a reactive metabolite RM. This in turn may be detoxified or react with cellular macromolecules to form covalently-bound genetic material (CBG). In this model, it is also possible that the reactive metabolite may be neutralized by nongenetic covalent binding (CBN). The covalently-bound genetic material may then be repaired (CBGR) or replicated (RCBG) resulting in the development of a genetic lesion.

Fig. 6.3 A more complex pharmacokinetic model for metabolic fate of a compound (from Gehring & Blau, 1977). C, chemical; RM, reactive metabolite; Ce, excreted chemical; IM, inactive metabolite; CBN, covalent binding, nongenetic; CBG, covalent binding, genetic; CBGR, repaired covalently bound genetic material; RCBG, retained genetic programme, critical and noncritical

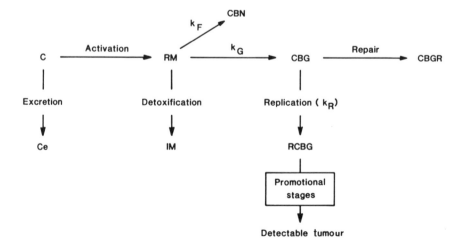

This model was subsequently examined in detail by Hoel *et al.* (1983). They assumed that all reactions are governed by linear first-order kinetics except for activation, detoxification and repair, which are allowed to be saturable in accordance with Michaelis–Menten kinetics. Since replication follows linear kinetics, the amount of damage is proportional to [CBG], the concentration of CBG. Under this model, [CBG] provides a measure of the effective dose.

Assuming that the dosing regimen is such that the concentration of a chemical [C] is proportional to the administered dose, one can use the two nonlinear differential equations describing the system first to solve for the concentration of a reactive metabolite [RM] in terms of [C], in steady state, and then to express [CBG] in terms of [RM].

Considering F to be of the one-hit form (6.18), Hoel *et al.* (1983) (see also Anderson *et al.*, 1980) noted that, under this model, the overall dose-response curve could assume any one of the four shapes shown in Figure 6.4, depending on which of the activation steps are saturable. With all processes being essentially linear, a linear dose-response curve results. If only repair or detoxification is saturable, the dose-response curve will be 'hockey-stick' shaped. If only activation is saturable, the shape is similar in form, although inverted. If more than one process is saturable, a combination of these shapes will occur.

Under this more complex model, the explicit expression for the overall dose-response model $P(d)$, as in (6.20), is very complicated. For the purposes of describing

Fig. 6.4   Relationship between delivered and administered dose and different pharmacokinetic conditions (from Hoel *et al.*, 1983)

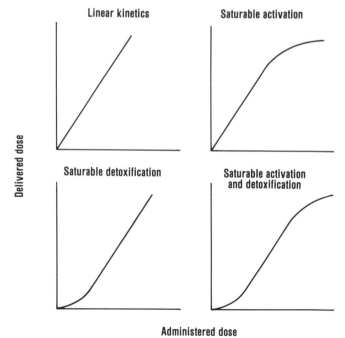

Linear kinetics   Saturable activation

Saturable detoxification   Saturable activation and detoxification

Delivered dose

Administered dose

actual data, the model in (6.22) may be fitted, however, using standard maximum likelihood procedures, as described below.

Maximum likelihood estimation

Whatever the specific parametric form of a dose-response model, the probability of observing a response at dose level d depends on some unknown parameters. In general, there may be p such parameters $\theta_1, \ldots, \theta_p$, summarized as a vector $\boldsymbol{\theta} = (\theta_1, \ldots, \theta_p)'$, the probability of response being then denoted as $P^*(d; \boldsymbol{\theta})$. Subsequently, we shall outline how these unknown parameters can be estimated from the observed data by maximum likelihood methods. Suppose that a total of n animals are used in an experiment involving $I + 1$ dose levels $0 = d_0 < d_1 < \cdots < d_I$ and that x_i of the n_i animals at dose d_i $(i = 0, 1, \ldots, I)$ develop tumours in the course of the study. Assuming that each animal responds independently of all other animals in the experiment, we find the likelihood of the observed outcome under any dose-response model $P^*(d; \boldsymbol{\theta})$ is given by

$$L(\boldsymbol{\theta}) = \prod_{i=0}^{I} \binom{n_i}{x_i} (P_i^*)^{x_i} (1 - P_i^*)^{n_i - x_i}, \tag{6.23}$$

where $P_i^* = P^*(d_i; \boldsymbol{\theta})$. Those values $\hat{\boldsymbol{\theta}}$ of the parameters $\boldsymbol{\theta}$ which maximize $L(\boldsymbol{\theta})$ are called 'maximum likelihood estimates'. Since maximization of $L(\boldsymbol{\theta})$ using direct analytical procedures is generally not possible, the maximum likelihood estimator $\hat{\boldsymbol{\theta}}$ of $\boldsymbol{\theta}$ is usually obtained using iterative numerical procedures. Under mild regularity conditions, it can be shown theoretically that $\hat{\boldsymbol{\theta}}$ is a consistent estimate for $\boldsymbol{\theta}$ as $n \to \infty$ (Krewski & Van Ryzin, 1981). Under these same conditions, $\sqrt{n}(\hat{\boldsymbol{\theta}} - \boldsymbol{\theta})$ is approximately normally distributed with mean $\mathbf{0}$ and variance/covariance matrix $\mathbf{V} = [(v_{rs})]$. The elements of the inverse of this covariance matrix $\mathbf{V}^{-1} = [(v^{rs})]$ represent the Fisher information and are derived through the second derivatives of the likelihood function $L(\boldsymbol{\theta})$ in (6.23):

$$v^{rs} = \sum_{i=0}^{I} c_i \frac{\partial P_i^*}{\partial \theta_r} \frac{\partial P_i^*}{\partial \theta_s} \bigg/ (P_i^* Q_i^*), \qquad (r, s = 1, \ldots, p), \tag{6.24}$$

where $c_i = \lim_{n \to \infty} n_i/n > 0$ and $Q_i^* = 1 - P_i^*$.

These theoretical results provide the basis for computing the estimated dose-response model $\hat{P}(d) = P^*(d; \hat{\boldsymbol{\theta}})$ and, when using the estimated covariance matrix of the parameter estimates, to calculate corresponding confidence intervals. For this purpose, the matrix $[(v^{rs})]$ is computed at $\boldsymbol{\theta} = \hat{\boldsymbol{\theta}}$, the actual values of the maximum likelihood estimates, and using $c_i = n_i/n$. The usual chi-square statistic

$$\chi^2 = \sum_{i=0}^{I} (x_i - n_i \hat{P}_i^*)^2 / (n_i \hat{P}_i^* \hat{Q}_i^*) \tag{6.25}$$

may be used to assess the goodness-of-fit of $\hat{P}^*(d)$. Provided that the assumed model $P^*(d)$ is correct, the asymptotic distribution of this statistic is chi-square with $(I + 1) - t > 0$ degrees of freedom.

Estimation procedures for the multistage model (6.13) are complicated by the non-negativity constraints on the parameters b_i. Because of these constraints, the asymptotic distribution of the maximum likelihood estimators will generally not be normal (Guess & Crump, 1978). Similarly, the usual chi-square statistic (6.25) will generally be inapplicable. Nonetheless, efficient algorithms for obtaining the restricted maximum likelihood estimators have been developed by Crump (1984) and Hartley and Sielken (1978).

As an example, consider the following data on liver tumours induced in mice in a lifetime study of feeding dieldrin (Walker *et al.*, 1973).

Dose (ppm):	0	1.25	2.50	5.00
Response (x_i/n_i):	17/156	11/60	25/58	44/60

These data, along with the fitted extreme value model, assuming independent background, are shown in Figure 6.5. The maximum likelihood estimates of the parameters ($\pm$ standard error) are $\hat{\alpha} = -2.46 \pm 0.50$, $\hat{\beta} = 1.66 \pm 0.35$ and $\hat{\pi}_0 = 0.106 \pm 0.024$. As may be expected with most monotonically increasing data sets, the model fits the data reasonably well, with no evidence of lack-of-fit provided by the chi-square statistic (p-value > 0.4). Similar results may be obtained with other data sets (Fig. 6.6 and Table 6.1) and with the other models discussed above. The fitted dose-response curves assuming additive background will also be similar, although in this case the likelihood surface is generally quite flat, and the parameters are thus less well determined.

Fig. 6.5 Dose-response curve for dieldrin-induced liver tumours in mice fitted under the extreme value model

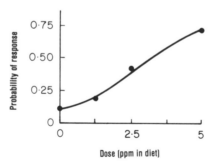

Estimation of quantiles

In some applications, interest centres on the added risk over background which, under the assumption of independence (6.8), would be

$$\Pi(d) = \frac{P^*(d) - P^*(0)}{1 - P^*(0)}. \tag{6.26}$$

Fig. 6.6   Dose-response curves for eight compounds fitted under the extreme value model

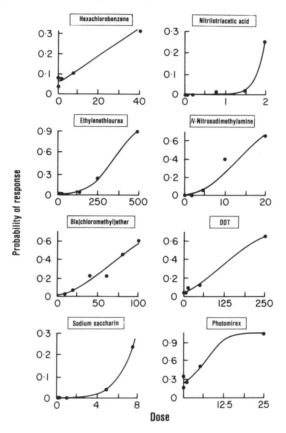

Table 6.1   Lesions induced by eight rodent carcinogens

Compound	Reference	Species	Tumour	Duration	Dose units
Hexachlorobenzene	Arnold *et al.* (1985)	Rat	Phaeochromo-cytoma	2 years	ppm in diet
Nitrilotriacetic acid	Food Safety Council (1978)	Rat	Kidney	2 years	ppm in diet
Ethylenethiourea	Graham *et al.* (1975)	Rat	Thyroid	2 years	ppm
N-Nitroso-dimethylamine	Terracini *et al.* (1967)	Rat	Liver	120 weeks	ppm
Bis(chloromethyl)-ether	Kuschner *et al.* (1975)	Rat	Respiratory	lifetime	no. of exposures
DDT	Tomatis *et al.* (1972)	Mouse	Liver	130 weeks	ppm
Sodium saccharin	Taylor & Friedman (1974)	Rat	Bladder	2 years	% in diet
Photomirex	Chu *et al.* (1981)	Rat	Thyroid	21 months	ppm in diet

In the same way, one may wish to evaluate the dose level corresponding to a certain level of risk over background, q, say $(0 < q < 1)$, which would be $d_q = \Pi^{-1}(q)$. The maximum likelihood estimator of d_q is defined by $\hat{d}_q = \hat{\Pi}^{-1}(q)$, where $\hat{\Pi}(d) = [\hat{P}^*(d) - \hat{P}^*(0)]/[1 - \hat{P}^*(0)]$. Since $\sqrt{n}\,(\hat{d}_q - d_q)$ is asymptotically normally distributed with mean zero and variance

$$\sigma^2 = \left\{ \frac{\partial \Pi}{\partial d}\bigg|_{d=d_q} \right\}^{-2} \sum_{r=1}^{P} \sum_{s=1}^{P} \frac{\partial \Pi}{\partial \theta_r} \frac{\partial \Pi}{\partial \theta_s} v_{rs}, \tag{6.27}$$

an approximate $100(1 - \alpha)\%$ confidence interval for d_q is given by

$$\hat{d}_q \pm z_{\alpha/2}\hat{\sigma}/\sqrt{n}, \tag{6.28}$$

where $z_{\alpha/2}$ denotes the $100(1 - \alpha/2)$ percentile of the standard normal distribution, and $\hat{\sigma}$ is an estimate of σ obtained by replacing $\boldsymbol{\theta}$ by $\hat{\boldsymbol{\theta}}$ in (6.27). Other possible confidence limit procedures, including those based on the asymptotic distribution of the log-likelihood (Cox & Hinkley, 1974, p. 343), are reviewed by Crump and Howe (1985).

Two applications of doses associated with certain levels of risk over background are of particular interest. First, Mantel and Bryan (1961) proposed the use of some suitably low risks (for example, $q = 10^{-6}$) as a means of defining a 'virtually safe' level of exposure in the absence of a threshold in the dose-response curve. It is now widely recognized that estimation of such extreme risks is subject to considerable model uncertainty. Extrapolation of the data on 2-AAF-induced liver tumours shown in Figure 6.7 (Littlefield et al., 1980a), for example, using the probit, logit, extreme value, multihit and multistage models, yields estimates of virtually safe doses spanning several orders of magnitude. Because of this uncertainty, it has been proposed that some form of linear extrapolation be used to obtain a lower limit on such extreme doses. The assumption of low-dose linearity follows, in fact, immediately from the assumption of additive background, since in that case a simple Taylor expansion shows

Fig. 6.7   Dose-response curve for 2-acetylaminofluorene(2-AAF)-induced liver tumours in mice fitted under the Weibull model and performance of six extrapolation procedures (from Krewski et al., 1984b). X, linear extrapolation; M, multistage model; W, Weibull model; L, logit model; G, gamma multi-hit model; P, probit model

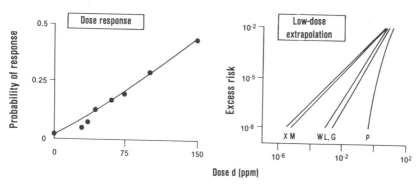

that

$$\Pi(d) \simeq f(0)d \qquad (6.29)$$

for small d, where $f(d) = \partial \Pi(d)/\partial d$. One simple procedure which may be used for this purpose is to extrapolate linearly from some higher quantile such as $d_{0.01}$ (Van Ryzin, 1980). (For the 2-AAF data, this form of linear extrapolation yields results close to those predicted by the multistage model.) A similar form of linear extrapolation has also been proposed by the World Health Organization (1984, p. 50).

The second application of the above concept, which can also be termed as estimating certain quantiles of the dose-response curve, is to derive a measure of carcinogenic potency. The measure proposed by Clayson *et al.* (1983) is defined by

$$C_q = K - \log_{10} d_q, \qquad (6.30)$$

where the dose d is expressed in μmol/kg body weight/day. The logarithm of dose is used to put the index on an order-of-magnitude basis, with the minus sign associating large values of C_q with low values of d_q. The constant K is set equal to 7 in order that C_q will usually lie in the range 1–10. By choosing a moderate value of q, say $0.10 \leq q \leq 0.50$, the model dependency encountered in estimating lower quantiles is avoided.

Despite its simplicity, the index C_q seems to provide a useful method of ranking animal carcinogens. Values of C_q with $q = 0.25$ for a selection of suspected and well-known animal carcinogens are shown in Table 6.2. Saccharin, the carcinogenicity of which has been widely debated, is assigned a potency index of 1.8, whereas the highly potent 2,3,7,8-tetrachlorodibenzo-*para*-dioxin (TCDD) has a value of 9.1. For a more complex ranking system that takes into account other factors, such as the spectrum of neoplasia induced and genotoxicity, the reader is referred to Squire (1981) and the related discussion by Crump (1983) and Theiss (1983).

Other quantitative measures of carcinogenic potency have also been proposed. Historically, Twort and Twort (1930, 1933) and Iball (1939) proposed several measures of potency in an attempt to summarize the data obtained from their experimental studies. For example, one of their measures was based on the time at which 25% of the

Table 6.2   Potencies $C_{0.25}$ of some selected compounds

Compound (Reference)	Species	Site	Potency $C_{0.25}$
Saccharin (Scientific Review Panel, 1983)	Rat	Bladder	1.8
2-Acetylaminofluorine (Littlefield *et al.*, 1980a)	Mouse	Bladder	4.3
		Liver	4.4
DDT (Thorpe & Walker, 1973)	Mouse	Liver	5.0
Aflatoxin (Wogan *et al.*, 1974)	Rat	Liver	8.6
Dioxin (National Toxicology Program, 1982)	Rat	Thyroid	9.1

animals would have developed tumours. Irwin and Goodman (1946) subsequently considered other related measures of potency, and noted that the different indices tended to give somewhat similar results. Bryan and Shimkin (1943) suggested the use of the dose required to induce tumours in 50% of the exposed animals, the quantity on which Meselson and Russel's (1977) index is defined.

More recently, Jones et al. (1981) considered the use of the time until 50% of the exposed animals would die from the tumour of interest as a measure of potency. Crouch and Wilson (1981) took the slope parameters in the one-hit model in (6.7) as a measure of potency. Noting that the probability of tumour induction $P(d) \simeq \lambda d$ for low levels of exposure d, Crouch and Wilson also proposed using the value of λ as a means of estimating the response rate P at a given dose d. This will be reasonable when the one-hit model, which is essentially linear even at moderate doses, provides an adequate description of the observed dose-response curve, but will be less satisfactory when the dose-response curve is highly nonlinear.

Sawyer et al. (1984) proposed a potency index based on the TD_{50}, or dose estimated to induce tumours in 50% of the exposed subjects. The method provides for the effects of both dose and time, and, like the method of Crouch and Wilson, is based on a simple exponential or one-hit model for the effects of dose. In cases where the time to tumour occurrence may not be directly observable, moreover, Sawyer et al. used the time to death with tumour present as a surrogate for the observable failure time (see Peto et al., 1984, for further discussion of this point).

Fig. 6.8   Range of carcinogenic potency in male rats (from Gold et al., 1984)

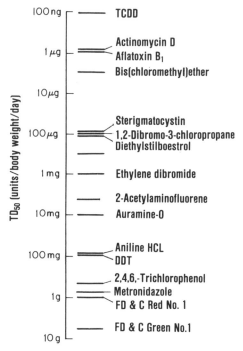

The index proposed by Sawyer *et al.* (1984) has recently been calculated using an extensive data base of known animal carcinogens compiled by Gold *et al.* (1984). Expressing the TD_{50} in terms of the daily intake of the compound relative to total body weight, this analysis revealed potencies varying over more than seven orders of magnitude (Fig. 6.8). Only a few nanograms of TCDD, for example, were required to induce a 50% tumour occurrence rate during the course of a rodent lifetime, whereas several grams of the food colours FD & C Red No. 1 and FD & C Green No. 1 were required to elicit the same rates of response.

6.3   Time-to-tumour models

Why use time-to-tumour models?

There are a number of situations in which the use of models based on time-to-tumour information has advantages. For example, when survival differs very greatly in different groups, one may be able to make fuller use of the data. In the extreme case, where near the end of an experiment there are survivors in only one group, most of the methods described so far cannot make use of any tumours detected in that group, as there is nothing with which to compare them. Where a parametric time-to-tumour model can be fitted, however, these data can be used to make more precise group comparisons.

Time-to-tumour models may allow one to present the results fully in a concise manner, which allows direct comparison with results from other comparable experiments. As we shall see, it is often possible to fit parametric models in which certain parameters, common to all groups, describe the general shape of the tumour incidence or survival curve, while a further single parameter, estimated separately for each group, describes the strength of the treatment effects. If common shape parameters are used, the strength parameters can be used to compare directly the results of different experiments, which would not be possible for methods based on testing a null hypothesis.

Graphical presentation of the observed and fitted time curves for tumour onset in treatment and control groups can be used to indicate whether there might be any interaction between the effect of treatment and time that should be investigated in more detail. The null hypothesis methods may, for example, not pick up a situation in which treatment increases tumour incidence early on in the study but decreases it later.

Time-to-tumour models are also of particular use in experiments specifically aimed at characterizing the mode of action of the carcinogen being tested. The shape of the time-response curve may assist in indicating whether the animal model used is apposite to the human situation where information on time-response may also be available. Furthermore, especially in more complex designs such as stopping experiments, it may give insight into whether the carcinogen had initiating or promoting action.

Finally, as noted in the previous section, time-to-tumour models may allow a more reliable method of low-dose extrapolation than those based on percentages of animals with tumours, especially where high doses markedly reduce mortality.

There are two main disadvantages of time-to-tumour models. One is the need to make the additional assumption, as compared with nonparametric methods, that tumour incidence follows a particular parametric relationship with time. A poor choice of relationship can affect conclusions as to the carcinogenicity of the treatment under test. The second is that, generally, far more extensive computing is required.

A general formulation

In the past, time-to-tumour models have mainly been applied to visible tumours. However, in recent years a number of attempts have been made to consider the more general situation where tumours may also be fatal or incidental (for example, Hartley & Sielken, 1977; Kodell *et al.*, 1982a; Peto *et al.*, 1984).

A simple way to model a long-term animal experiment is illustrated in Figure 6.9, where the possible states an animal can be in are given as boxes, and the connecting arrows indicate possible transitions (Kodell & Nelson, 1980). In this model, an animal starts in a normal disease-free state (N) and may at some time either develop a tumour (T) – or perhaps more precisely be in a state where a tumour can be detected – or die from a cause unrelated to the tumour of interest (D_{NT}). An animal with a tumour may also subsequently die from this unrelated cause, or may die because of the tumour (D_T).

Fig. 6.9   Illustration of illness and death states with possible transitions in rodent bioassay; *N*, normal; *T*, tumour; D_T, death from tumour; D_{NT}, death not from tumour (from Kodell & Nelson, 1980)

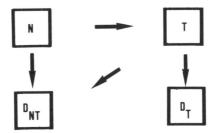

This model is aimed at the types of observation which can arise from a long-term experiment. The assumption that a transition is made from a normal state to a state where the tumour is detectable is simplistic, inasmuch as it does not take into account details of the underlying biological processes. However, it would not in practice be possible to identify all of these states, even by increasing the number of investigations. Even in the model proposed, the information on cause of death, required to distinguish between D_T and D_{NT} when a tumour is present, is often difficult or impossible to obtain.

Three random variables may be used to describe the above model:

(a) X: time to onset of tumour, or transition time from the normal state (N) to the tumour-bearing state (T)

(b) Y: time to death due to tumour, or transition time from the normal state (N) to the death-from-tumour state (D_T)

(c) Z: time to death from an unrelated cause, or transition time from the normal state (N) to the death-not-from-tumour state (D_{NT}).

The random variable Z is not of major concern in drawing statistical conclusions on the process of carcinogenesis and is not considered further, except with regard to assumptions about Z needed to form the likelihood functions upon which statistical inferences may be based. The two random variables X and Y, however, which have to satisfy the condition $X \le Y$, are of major interest for the statistical inference.

Let $G_X(t)$ and $G_Y(t)$ be the distribution functions of X and Y, or $S_X(t) = 1 - G_X(t) = \mathrm{pr}(X \ge t)$ and $S_Y(t) = 1 - G_Y(t) = \mathrm{pr}(Y \ge t)$ the survivor functions of X and Y.

In addition, we consider for any index X or Y the density $f(t) = dG(t)/dt$, the hazard function $\lambda(t) = f(t)/S(t)$ and the cumulative hazard function, $\Lambda(t) = \int_0^t \lambda(u)\, du$. Note that $S(t) = \exp[-\Lambda(t)]$.

The hazard function $\lambda(t)$ is an extremely useful tool for modelling distributions of random variables which represent the time to a well defined event. It can also be defined as the conditional probability that the event of interest occurs at time t, if it has not occurred before that time. Let X be the random variable of interest:

$$\lambda_X(t) = \lim_{\Delta t \to 0} \frac{\mathrm{pr}(t \le X < t + \Delta t \mid X \ge t)}{\Delta t} = \frac{f_X(t)}{S_X(t)}.$$

The hazard function is therefore also denoted the age-specific failure rate.

In a long-term animal experiment, considering visible or occult tumours, four different events can be observed at a certain time point:

A. appearance of a visible tumour
B. death caused by the tumour of interest (fatal context)
C. death from unrelated cause, tumour of interest present (incidental context)
D. death without tumour of interest.

It has to be noted that deaths under C and D, from a cause unrelated to the tumour of interest, can occur in scheduled sacrifices. Thus, animals killed because of the experimental design or other external reasons represent observations of type C and D, depending on whether the tumour of interest is seen or not seen in the particular animal.

The contribution to the likelihood function of the four types of observations, expressed in terms of the two random variables X and Y, are as follows:

A. density of X: $f_X(t)$
B. density of Y: $f_Y(t)$
C. $\mathrm{pr}(X < t \le Y)$: $S_Y(t) - S_X(t)$
D. survivor function of X: $S_X(t)$.

In a given experiment, let $t_1, t_2, \ldots, t_K$ be the distinct times at which events of the above type are observed; a_k, b_k, c_k and d_k ($k = 1, \ldots, K$) are the number of events of type A, B, C or D at time t_k. The latent-failure-time approach is taken in the formation of likelihoods below. Although this is the most common approach to the competing risks problem, strong yet unverifiable assumptions are required to form the likelihoods for occult tumour data (Kalbfleisch *et al.*, 1983).

We consider four typical types of experiments:

(i) *Visible tumours*

As the time of onset is observable directly, there is no interest in the random variable Y; only events of the type A and D are observed, leading to a likelihood function

$$L_1 = \prod_{k=1}^{K} \{f_X(t_k)\}^{a_k} \{S_X(t_k)\}^{d_k}.$$

With $f_X(t) = \lambda_X(t)S_X(t)$ and $S_X(t) = \exp(-\Lambda_X(t))$, the log-likelihood function is often expressed as

$$LL_1 = \sum_{k=1}^{K} \{a_k \log[\lambda_X(t_k)] - (a_k + d_k)\Lambda_X(t_k)\},$$

using only the hazard function and its cumulative version, which are often the basic entities on which parametric models are formulated.

(ii) *Occult tumours – all tumours observed in a fatal context*

In this special situation, where no tumours are found in animals dying from unrelated causes, only events of types B and D are observed. The likelihood function is

$$L_2 = \prod_{k=1}^{K} \{f_Y(t_k)\}^{b_k} \{S_Y(t_k)\}^{d_k}$$

or, as above, the log-likelihood function

$$LL_2 = \sum_{k=1}^{K} \{b_k \log[\lambda_Y(t_k)] - (b_k + d_k)\Lambda_Y(t_k)\},$$

which are formally identical to L_1 and LL_1, respectively.

(iii) *Occult tumours – all tumours observed in an incidental context*

In this case $S_Y(t) = 1$, as no deaths due to the tumour of interest are occurring, only events of the types C and D being observed. In order to form a likelihood for this case it must be assumed that tumour-bearing and tumour-free animals of the same age have identical hazard functions for death unrelated to tumour (that is, intercurrent mortality). The likelihood function under this assumption is

$$L_3 = \prod_{k=1}^{K} \{1 - S_X(t_k)\}^{c_k} \{S_X(t_k)\}^{d_k}.$$

The log-likelihood function expressed in terms of $\Lambda_X(t)$ would be

$$LL_3 = \sum_{k=1}^{K} \{c_k \log[1 - \exp(-\Lambda_X(t_k)] - d_k\Lambda_X(t_k)\}.$$

(iv) *Occult tumours – observed in fatal or incidental context*

In this most general case, events of the types B, C and D are observed. In order to form a likelihood for this case, it must, of course, be assumed that each tumour can be reliably classified as either fatal or incidental. In addition, it must be assumed that tumour-bearing and tumour-free animals of the same age have identical hazard functions for death unrelated to tumour. The likelihood function under these assumptions is

$$L_4 = \prod_{k=1}^{K} \{f_Y(t_k)\}^{b_k}\{S_Y(t_k) - S_X(t_k)\}^{c_k}\{S_X(t_k)\}^{d_k}.$$

Kodell *et al.* (1982a) pointed out that

$$S_Y(t) - S_X(t) = [1 - Q(t)]S_Y(t),$$

with $Q(t) = S_X(t)/S_Y(t) = pr(X > t \mid Y > t)$ being the conditional probability of tumour onset after time t, given tumour-free survival through time t. Even under the strong assumptions noted above, the function $S_Y(t)$ does not have a simple interpretation in terms of time to death due to tumour alone (that is, independent of the influence of time to onset of tumour).

From this it follows that

$$L_4 = \prod_{k=1}^{K} \{f_Y(t_k)\}^{b_k}\{S_Y(t_k)\}^{c_k+d_k}\{1 - Q(t_k)\}^{c_k}\{Q(t_k)\}^{d_k},$$

which can be written as the product of two parts. One,

$$L_4^{(1)} = \prod_{k=1}^{K} \{f_Y(t_k)\}^{b_k}\{S_Y(t_k)\}^{c_k+d_k},$$

depends on $S_Y(t)$ only; the other,

$$L_4^{(2)} = \prod_{k=1}^{K} \{1 - Q(t_k)\}^{c_k}\{Q(t_k)\}^{d_k},$$

on $Q(t)$ only.

The two components of the log-likelihood function are

$$LL_4^{(1)} = \sum_{k=1}^{K} \{b_k \log[\lambda_Y(t_k)] - (b_k + c_k + d_k)\Lambda_Y(t_k)\},$$

which is similar to LL_2 above; and, taking $\Lambda_Q(t) = -\log Q(t)$, the second part

$$LL_4^{(2)} = \sum_{k=1}^{K} \{c_k \log[1 - \exp(-\Lambda_Q(t_k)] - d_k\Lambda_Q(t_k)\},$$

which is similar to LL_3.

In the case where all tumours are fatal or visible, nonparametric estimation of the survivor functions can be made by the Kaplan–Meier method, as discussed in Section 5.3. Where all tumours are incidental, nonparametric estimation can be done by the method of Hoel and Walburg, as discussed in Section 5.5. Kodell *et al.* (1982a), who consider the fourth, most general case where tumours may be either fatal or incidental, note that the Kaplan–Meier estimator can be applied to the first part, $L_4^{(1)}$ or $LL_4^{(1)}$, to estimate $S_Y(t)$. With the assumption that the ratio $Q(t)$ is monotonically decreasing, they then derive a nonparametric maximum likelihood estimation for $S_X(t)$ as well.

Dinse and Lagakos (1982) weaken the restriction of monotonicity of $Q(t)$ and propose to estimate $S_X(t)$ in the class of all survivor functions that are stochastically smaller than $S_Y(t)$. The Kaplan–Meier estimate of $S_Y(t)$ is used as a starting point in an iterative approach. Given this estimate, $LL_4^{(2)}$ is maximized for $Q(t)$ under the restriction that the resulting estimate must be a monotone decreasing survival function. This is enforced by applying techniques of isotonic regression. The estimate for $S_X(t)$ thus derived is then inserted into the original likelihood function, which is maximized again for $S_Y(t)$. This process is iterated until convergence.

Turnbull and Mitchell (1984) have proposed a simpler algorithm, by addressing the problem in terms of the joint distribution of time of onset of tumour (X) and time of death from tumour (Y), which have marginal distributions $1 - S_X(t)$ and $1 - S_Y(t)$. They point out that the joint distribution can have positive mass only on a finite set of disjoint intervals in the (x, y)-plane, which can be constructed from the observations. They then use the EM-algorithm to estimate the probability masses to be attributed to these intervals. The marginal distributions, and hence the survivor functions $S_X(t)$ and $S_Y(t)$, can then be derived from the estimate of the joint distribution.

Choice of parametric model

The ideal model to use would clearly be one derived from sound biological theories of the carcinogenic process. In practice, of course, understanding of the carcinogenic process is far from perfect, but one can still aim for a model which both fits in with available knowledge and fits observed data at least reasonably well. By far the most attention has been given to the multistage model and, in the case of carcinogenesis experiments in which the dose has been applied continuously or at regular intervals, to the use of the Weibull distribution derived from it. For this reason, and also because it has proved satisfactory for analysis of a considerable number of data sets, we will consider this in detail first and turn our attention to other models only at the end of this section.

Multistage models and derivation of the Weibull distribution

Armitage and Doll (1954) observed that, in humans, the age-specific incidence rates of many types of cancer are proportional to a power of age (or time from first exposure) and showed that this result would be expected under a multistage model. This model makes the following simplifying assumptions:

(i) that there is a large and constant number (N) of cells at risk,
(ii) that all the cells start in an identical state,

(iii) that at least one of them has to undergo a fixed number κ of stages or transformations before a tumour appears, and

(iv) that any cell in a given stage has the same very small, but constant probability per unit time of commencing the transformation into the next stage (kinetic rate-constants $\delta_1, \delta_2 \cdots \delta_\kappa$).

Under these assumptions, the number of cells that have undergone the first $\kappa - 1$ stages by time t is given by

$$N_{\kappa-1} = N \int_0^t \delta_{\kappa-1} \int_0^{u_{\kappa-1}} \delta_{\kappa-2} \cdots \int_0^{u_2} \delta_1 \, du_1 \, du_2 \cdots du_{\kappa-1} = \frac{N\delta_1\delta_2 \cdots \delta_{\kappa-1} t^{\kappa-1}}{(\kappa-1)!}.$$

If it is further assumed that each transformation takes a constant time $(w_1, w_2 \cdots w_\kappa)$, the formula for the incidence rate $I(t)$ becomes

$$I(t) = \beta\kappa(t - w)^{\kappa-1},$$

where $\beta = N\delta_1\delta_2 \cdots \delta_\kappa/(\kappa!)$ and $w = \sum_{j=1}^\kappa w_j$ is the sum of the constant times of the transformations, which can be seen as the minimal time to tumour. This is a particular way of parametrizing the Weibull distribution. The survivor function is then

$$S(t) = \exp(-\beta(t - w)^\kappa). \tag{6.31}$$

While this model is clearly an over-simplification, it may be expected that if cancer mechanisms in animals and humans are of this type, incidence rate data from laboratory animals, which are often inbred and presumably therefore more homogeneous, may follow a Weibull distribution even more clearly than for humans. The suggestion of using Weibull distributions to analyse continuous carcinogenesis experiments in animals where time-to-tumour is observable was first made by Pike (1966).

From a study of the way in which the formula was derived, it should be clear that κ and w are inherent properties of the process being studied and should not vary between groups within an experiment studying the same cancer type in the same species of animals. The parameter β, on the other hand, should be affected by treatment, assuming that the effect of continuous application of a carcinogen will be to alter at least one of the kinetic rate constants. In an experiment with several dose groups, $i = 1, \ldots, I$, say, one would consider different parameters $\beta_1, \ldots, \beta_I$ for the Weibull distribution of time to tumour for the animals from the respective groups, or, as will be outlined later, one would model the dependence of the β_i on the dose level or other covariates.

Fit of Weibull distributions to visible tumour data

Weibull distributions have been used mainly in the literature to analyse visible tumour data, often from mouse skin-painting experiments widely used in the 1960s and 1970s to evaluate the carcinogenicity of tobacco-smoke condensates and of other chemicals. In the next few sections we consider these applications in some detail, before turning to more recent work using Weibull distributions in the analysis of tumours not visible in life. Some of the ideas used in the visible tumour analyses (for example, for significance testing of treatment effects) have natural analogues for nonvisible tumours, but are described in detail only for the former situation.

Maximum likelihood estimation of the parameters β, κ and w of the Weibull distribution is discussed in detail by Peto and Lee (1973). The contribution of the log-likelihood LL_1 of an animal dying without a tumour at time t' is given by

$$\log[S_X(t')] = -\beta(t' - w)^\kappa,$$

while that of an animal getting a tumour at time t'' is given by

$$\log[\lambda_X(t'')S_X(t'')] = \log \beta + \log \kappa + (\kappa - 1)\log(t'' - w) - \beta(t'' - w)^\kappa.$$

To illustrate the fitting of Weibull models, consider the cigar-smoke condensate data (Section 4.2). The maximum likelihood estimates (and standard errors obtained from the inverse of the information matrix) of w, κ and β for the three cigar-smoke condensate groups are given in Table 6.3.

Table 6.3 Parameter estimates of Weibull models fitted to the three groups, from data on cigar-smoke condensate

	Low dose	Middle dose	High dose
$\hat{w} \pm SE$	19.5 ± 8.6	30.4 ± 10.1	34.5 ± 4.9
$\hat{\kappa} \pm SE$	1.7 ± 0.6	2.3 ± 0.8	2.0 ± 0.5
$(\hat{\beta} \pm SE) \times 10^4$	2.13 ± 5.95	0.62 ± 2.25	3.17 ± 6.61
Log-likelihood	-131.522	-164.084	-180.537

Because the estimates of κ and w have not been constrained to be equal in the three groups, the estimates in the table do not provide a particularly useful summary of the relationship between cigar-smoke condensate dose and tumour incidence. The standard error estimates are relatively large due to the fact that the likelihood functions are quite flat around their maxima. Accordingly, when making inferences about the Weibull parameters, likelihood ratio tests should be employed rather than tests based on the asymptotic normality of the maximum likelihood estimators.

Estimated percentiles often provide a useful summary of the Weibull time-to-tumour curves. Estimates of percentiles can be obtained directly from the estimated tumour incidence curves, and standard errors of estimated percentiles can be derived using the delta method (Miller, 1981b, pp. 25–27). Estimates of the 25th percentile (denoted T_{25}) and the 50th percentile (that is, the median, denoted T_{50}) for the three cigar-smoke condensate groups are as follows:

	Low dose	Middle dose	High dose
$T_{25} \pm SE$	84.5 ± 8.3	71.4 ± 4.2	62.5 ± 3.1
$T_{50} \pm SE$	127.6 ± 19.2	90.8 ± 4.8	77.5 ± 3.5

The percentiles provide a much more useful summary of the Weibull curves than was provided by the estimates of the three Weibull parameters. They indicate that the tumours are appearing earlier with increasing dose of cigar-smoke condensate. The standard errors of the percentile estimates, with the exception of the median for the low-dose group, are less than 10% of thei. corresponding estimates. The standard error for the estimated median of the low-dose group is large because the estimate, in this case, represents an extrapolation outside the observed time period.

To formulate the full log-likelihood function LL_1, in the situation of common values for w and κ but different parameters β_i ($i = 1, \ldots, I$) for the different experimental groups, we have to extend our notation slightly. Let a_{ki} and d_{ki} be the number of events of type A and D observed at time t_k in group i ($k = 1, \ldots, K; i = 1, \ldots, I$). Then we have

$$LL_1 = \sum_{k=1}^{K} \sum_{i=1}^{I} [a_{ki}\{\log \beta_i + \log \kappa + (\kappa - 1)\log(t_k - w)\} - (a_{ki} + d_{ki})\beta_i(t_k - w)^\kappa].$$

Maximization of this log-likelihood function, which depends on $\beta_1, \ldots, \beta_I$, κ and w, is not straightforward, but can be achieved satisfactorily by a modified Newton–Raphson iterative procedure.

For the cigar-smoke condensate data, the maximum likelihood estimates of $w(\pm SE)$ and $\kappa(\pm SE)$ are 17.5($\pm$6.45) and 2.8($\pm$0.5), respectively, and the maximum likelihood estimates of β_i are as follows:

	Low dose ($i = 1$)	Middle dose ($i = 2$)	High dose ($i = 3$)
$(\hat{\beta}_i \pm SE) \times 10^6$	2.28 ± 5.22	4.50 ± 10.22	7.69 ± 17.28

The parameter estimates now are indicative of an increasing tumour rate with increasing dose of cigar-smoke condensate. The log-likelihood for this model, with common w and κ but different β_i, is -481.173. Fitting a model with a common β in addition to common w and κ gives a log-likelihood of -491.353. Thus, the likelihood ratio test statistic for $H_0 : \beta_1 = \beta_2 = \beta_3$ is computed as $2(491.353 - 481.173) = 20.36$. Comparing this computed value to a table of percentiles for the chi-square distribution with two degrees of freedom (in general, $I - 1$ degrees of freedom) we find that $p = 0.00004$, indicating a strong effect of cigar-smoke condensate on tumour incidence.

The sum of the log-likelihoods from the initial fits of separate Weibull models to each dose group is $-131.522 - 164.084 - 180.537 = -476.143$. Thus, a test of the null hypothesis that $w_1 = w_2 = w_3$ and $\kappa_1 = \kappa_2 = \kappa_3$ can be based on the likelihood ratio test statistic, which is computed as $2(481.173 - 476.143) = 10.06$. Comparing this computed value to a table of percentiles for the chi-square distribution with four degrees of freedom (in general, $2I - 2$ degrees of freedom), we find that $p = 0.039$, indicating some evidence of heterogeneity. It is interesting to note that in spite of this evidence that the assumption of common w and κ may be invalid, the likelihood ratio test statistic for $H_0 : \beta_1 = \beta_2 = \beta_3$ differed only slightly from the log-rank test statistic calculated for the cigar-smoke condensate data in Section 5.6 (that is, $X^2_{LH} = 20.16$).

Peto and Lee (1973) discuss various possibilities for proceeding in the case of known or unknown parameters κ and β. Of particular importance is the situation in which the main interest is in between-treatment comparison, that is, in the relative magnitude of the β_i values. Then, rather than carry out full estimation of κ, w and β_i, it is reasonably satisfactory, and much simpler computationally, to fix κ and w values from previous experience and compute the β_i from the formula

$$\beta_i = s_i/v_i,$$

where $s_i = \sum_{k=1}^{K} a_{ki}$ is the number of animals bearing tumours in the ith group and $v_i = \sum_{k=1}^{K} (t_k - w)^{\kappa}$, the summation being over both times of tumour and times of death without tumour in group i. Pike (1966) has demonstrated that the ratio of β's between different groups is virtually independent of the actual values of κ and w chosen, provided the (κ, w) pair is not too far from the best fitted values. Where κ and w are known, the asymptotic variance of β_i is given by var $\beta_i = \beta_i^2/s_i$.

Goodness-of-fit to the Weibull distribution can be tested by dividing the experimental time period into J intervals $(T_0 = 0, T_1]$, $(T_1, T_2], \ldots, (T_{J-1}, T_J]$. Within each interval one compares the observed number of animals in the ith group first developing a tumour in the jth interval O_{ij} with the number expected E_{ij}. E_{ij} is calculated by $E_{ij} = \beta_i v_{ij}$, where j refers to the interval $(j = 1, \ldots, J)$, and v_{ij} is calculated by summing, for each animal of group i surviving and tumour-free at T_{j-1}, the term $(t^* - w)^{\kappa} - (T_{j-1} - w)^{\kappa}$, where $t^* = \min(T_j, t_k)$, that is, the time to death or tumour for animals that experience one of these events during the jth interval, or the upper limit of the interval for the remainder. If the numbers of tumours are too small per group, it will often be useful to combine these O_{ij} and E_{ij} values over groups for each time interval:

$$O_j = \sum_{i=1}^{I} O_{ij} \quad \text{and} \quad E_j = \sum_{i=1}^{I} E_{ij}.$$

The statistic $X^2 = \sum_{j=1}^{J} (O_j - E_j)^2/E_j$ should then produce an approximate chi-square variable on $J - 1$ or $J - 3$ degrees of freedom, depending on whether κ and w were assumed to be known or were fitted from the data.

Treatment effects: estimation and significance testing

If κ and w are known or have been estimated from the data, then the log-likelihood for an I group experiment is given by

$$LL = \sum_{i=1}^{I} s_i \log \beta_i - \sum_{i=1}^{I} \beta_i v_i.$$

If the parameters β_i for each group depend on certain covariates as explanatory variables (dose, carcinogen, method of application, etc) $z_1 \cdots z_p$, where z_{iu} is the value of the uth variable in the ith group, then it is convenient to relate β_i to the z_{iu} by the expression

$$\log \beta_i = \sum_{u=1}^{p} \theta_u z_{iu}$$

or

$$\beta_i = \exp \sum_{u=1}^{p} \theta_u z_{iu},$$

where the θ_u are regression coefficients to be estimated. It can be said that a log-linear model for the β_i is used. The log-likelihood function then becomes

$$LL = \sum_{i=1}^{I} s_i \sum_{u=1}^{p} \theta_u z_{iu} - \sum_{i=1}^{I} v_i \exp \sum_{u=1}^{p} \theta_u z_{iu}.$$

Multiple regression methods based on maximum likelihood estimation are used for this problem. Likelihood ratio tests can be employed to investigate the significance of certain covariates. Let $LL^{(1)}$ be the log likelihood of a given model with a certain number of covariates. Inclusion of d further covariates alters the fitted log likelihood to $LL^{(2)}$. Under the null hypothesis that the regression coefficients θ_u for the newly included covariates are zero, $2(LL^{(1)} - LL^{(2)})$ should be approximately chi-square-distributed with d degrees of freedom. For detailed illustrations, see Peto and Lee (1973).

Aitkin and Clayton (1980) have published a computer program to fit such regression models to possibly censored failure-time data as they arise in this context. Their program is developed in the framework of the GLIM package (Baker & Nelder, 1978) for fitting generalized linear models.

Support for the model

The fourth assumption of the multistage hypothesis underlying the Weibull distribution implies that, if treatment is continuous, the kinetic rate-constants for each stage do not depend on the age of the animal. It follows that, provided the carcinogen affects the first stage of the process strongly enough for it to be a reasonable approximation to assume all observed tumours to have arisen because of this, the age-specific incidence rates will depend wholly on the duration of treatment and not at all on age *per se*.

In a large experiment carried out by Peto *et al.* (1975), 3,4-benzo[*a*]pyrene (BP) was applied to the skin of mice in four groups of increasing size starting at 10, 25, 40 and 55 weeks old, respectively. As can be seen from Figure 6.10, the percentage of mice without a tumour, when plotted against age, differed markedly between the four groups. However, when plotted against treatment duration, the four groups were virtually identical, as expected from multistage assumptions. Having shown that the relationships between tumour incidence and duration in the four groups were not significantly different, Peto *et al.* (1975) combined the results of the four groups to illustrate the overall fit to the Weibull distribution. This is shown in Figure 6.11; it is obvious that, over the 100-fold range of incidences from 0.25% per fortnight up to the massive rate of 25% per fortnight observed after 90 weeks of regular BP administration, the points did approximately fit the theoretical straight line obtained by taking logarithms of the Weibull equation $I = \beta(t - w)^\kappa$. The multistage model also predicts that, if a carcinogen has an effect directly proportional to dose on each of c (of κ) kinetic-rate constants, and if the dose applied is sufficiently large for the background

Fig. 6.10 Percentage of tumourless mice against (a) age or (b) duration of exposure to benzo[a]pyrene (from Peto *et al.,* 1975)

(a)

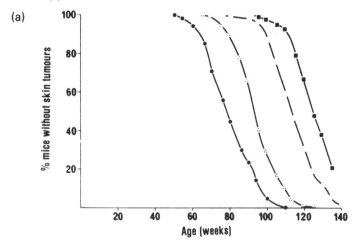

(b)

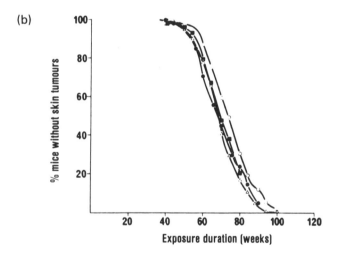

KEY

● = GROUP 1
△ = GROUP 2
○ = GROUP 3
■ = GROUP 4

rate-constants to be neglected for those c stages of the cancer process, the age-specific tumour incidence rate will then be proportional to dose to the power of c. Thus, if, say, stages two and three of a five-stage process are affected linearly by treatment, d is dose, δ_i are background rate-constants and ϕ_i is the increment in rate-constant per unit

Fig. 6.11   Incidence rates of 10-mm epithelial tumours at successive fortnightly chartings
           against duration of benzo[*a*]pyrene (BP) application, on a log-log scale from 28
           weeks onwards. (The points are statistically independent, and 90% confidence
           intervals are indicated.) (from Peto *et al.*, 1975)

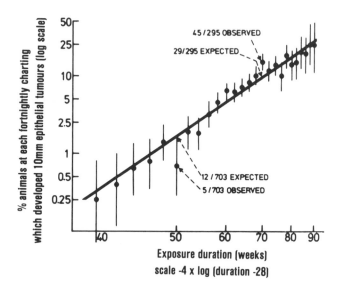

dose for affected stages, the Weibull parameter β will be proportional to

$$\delta_1(\delta_2 + \phi_2 d)(\delta_3 + \phi_3 d)\delta_4\delta_5.$$

As d becomes large, this approximates to

$$\delta_1\delta_4\delta_5\phi_2\phi_3 d^2.$$

Another way of testing whether the observed failure times comply with the Weibull
distribution is based on the fact that for the Weibull distribution $\log\log[1/S(t)] =
\log\beta + \kappa\log(t - w)$. Note that the left-hand side can also be written as $\log[-\log S(t)]$.
Using a nonparametric estimate of the survivor distribution $S(t)$ (Kaplan–Meier
estimate discussed in Section 5.3) and plotting its above transform against the
logarithm of time provides a simple check of the model.

Lee and O'Neill (1971) analysed an experiment in which BP was painted continuously at 6, 12, 24 and 48 μg per week on four groups of 300 mice. They found that
not only could skin tumour incidence be well described by a Weibull distribution with κ
and w common to all four groups, but that β was proportional to dose squared. As can
be seen in Figure 6.12, the plots of $\log\log[1/\hat{S}(t)]$ against $\log(t - 17.70)$ form
approximately parallel equidistant straight lines. The slope of the lines, 2.95, estimates
κ and is not significantly different from an integer value as suggested by the model. The
average vertical difference between the lines is almost exactly twice the logarithm of
the ratio of successive doses implying $c = 2$ and thus the results are consistent with a
multistage hypothesis in which BP affects two out of three of the stages of the process.

Fig. 6.12   Fit of Weibull distribution to data from a skin-painting experiment with
benzo[a]pyrene in mice (from Lee & O'Neill, 1971)

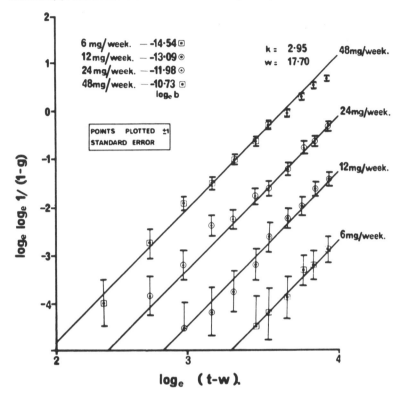

Lee *et al.* (1977) describe the analysis of a series of ten mouse skin painting experiments in which there were a total of 55 treatment groups consisting of either whole cigarette smoke condensate (SWS) or various fractions of it tested at varying dose levels. Common values of $\kappa = 3.05$ and $w = 11.29$ were fitted to the skin tumour data by maximum likelihood methods and the following linear model for the remaining Weibull parameter

$$\log \beta_{ij} = \mu + \alpha_i + q \, (\log \text{dose}_j)$$

was fitted to the responses for the ith treatment (fraction) and jth dose level. This approach leads to a simple description of the results in terms of the 'tumorigenic ratio' which measures the activity of the fraction relative to whole-smoke condensate on a weight-for-weight basis.

Druckrey (1967) reported quantitative dose-response relationships incorporating time to response information for a variety of chemical carcinogens and established the now well-known relationship

$$d \cdot t^n = \text{const.},$$

where d is the daily dose and t the median 'tumour induction time'. This empirical relationship can be seen as a corollary of the Weibull model (6.31) (Carlborg, 1981). Consider $w = 0$ and the parameter β being proportional to some power of the daily dose d, that is, $\beta = \alpha \cdot d^m$. The absence of a constant, not dose-dependent term in this submodel for β implies a zero background response. Solving (6.31) for the median induction time gives

$$0.5 = \exp(-\alpha d^m t^\kappa)$$

or

$$d t^n = \{(\log 2)/\alpha\}^{1/m} = \text{const.}$$

with $n = \kappa/m$.

Noncontinuous exposure

Although the examples considered above concerned only the analysis of experiments in which skin tumours were produced in mice by regular skin painting, a Weibull distribution has also been successfully fitted to experiments with rats in which a single intrapleural inoculation of asbestos resulted in mesotheliomas of the pleura (Berry & Wagner, 1969; Wagner *et al.*, 1973). There are two theoretical reasons why a Weibull distribution might fit in this situation. One is that, although the injection of asbestos is given as a single dose, the asbestos is not easily destructible and remains in the animal for a considerable time after injection, thus simulating continuous exposure. The second is that, in the multistage model, if the effect of a single exposure is so large that a substantial proportion of cells at risk are transformed very rapidly through the first κ^* stages, with a subsequent tumour occurring only after background transformations cause the remaining $\kappa - \kappa^*$ transformations, the incidence rate will still obey a Weibull distribution but with a parameter $\kappa - \kappa^*$ and not κ.

In other experiments, such as those described by Day and Brown (1980), animals have been exposed for varying lengths of time and then treatment has been stopped. It is clear that in many of these experiments a simple Weibull distribution does not fit the observed response. This is not surprising, as the fourth assumption of the multistage hypothesis will not hold since kinetic rate constants of the stages affected by treatment will presumably change on stopping treatment. A number of workers have considered the mathematical implication of the application of multistage models to 'stopping experiments' (Lee, 1975; Whittemore & Keller, 1978; Day & Brown, 1980; Parish, 1981). For example, in a three-stage model in which treatment causing kinetic rate constants α_1, α_2 and α_3 was applied up to time S and then stopped, causing reversion to background kinetic rate-constants δ_1, δ_2 and δ_3, the incidence rate at time T ($>S + w$), in the simplified situation where the waiting time w all occurs after the final transformation, is given by

$$N\left[\frac{\alpha_1\alpha_2\delta_3 S^2}{2} + \alpha_1\delta_2\delta_3 S(T - S - w) + \delta_1\delta_2\delta_3 \frac{(T - S - w)^2}{2}\right];$$

the above references give details of formulae in more general situations (κ stages rather than three; individual waiting times for each stage). It should be noted that the shape of the incidence curve with time after stopping depends on which stages it is assumed that the carcinogen affects. Thus, if only the first stage is affected ($\alpha_1 > \delta_1$, $\alpha_2 = \delta_2$, $\alpha_3 = \delta_3$), the fall off in incidence compared with continuous exposure will be much less pronounced than if later stages are affected. In particular, incidence will tend to be approximately constant for some time after stopping if the penultimate stage only is affected, and will tend to drop sharply to background levels after stopping if the final stage is affected. Lee (1975) has used maximum likelihood methods in an attempt to distinguish formally between hypotheses in which a carcinogen does and does not affect a certain stage or stages. However, the computation involved is considerable.

Further discussion of details of analysis of these special experimental situations is outside the scope of this monograph, although it is worth pointing out that the methods for stopping experiments can also be applied to crossover experiments in which varying treatments are given in varying orders to the same animals.

Fitting Weibull distributions to data for internal tumours

In their analysis of data from the British Industrial Biological Research Association (BIBRA) nitrosamine study, some of which are described in Section 4.3, Peto *et al.* (1984) successfully used related Weibull distributions to describe the distribution of time X to onset of tumour and of time Y to death because of tumour. Referring back to the general formulation given above, they assumed $\Lambda_X(t) = \beta t^\kappa$ and $\Lambda_Y(t) = \beta f t^\kappa$. The additional parameter f they referred to as the 'fatality factor', ranging from 0 for a completely nonfatal tumour to 1 for an instantly fatal tumour.

Over the wide range of dose levels tested, f (and κ) appeared to be essentially invariant of dose. This allowed characterization of the dose-response relationship for liver and oesophageal tumours in terms of a single parameter β for each tumour type. While more experience is needed with this model, it appears to be a very useful approach.

Kodell and Nelson (1980) use the Weibull distribution within their simplified observational model for the carcinogenic process, which was introduced above (Fig. 6.9). They consider the transition time from N to T, which corresponds to the random variable X for time to onset of tumour, to follow a Weibull distribution. Their parameterization of the hazard function is $\beta_1 t^{\gamma_1}$. They further consider the transition time from T to D_T, which corresponds to the random variable $Y - X$, being of a Weibull type with hazard function $\beta_2 t^{\gamma_2}$ as well as transition from N or T to D_{NT} with hazard function $\beta_3 t^{\gamma_3}$. Within this framework, the likelihood function is developed considering both natural deaths as well as scheduled sacrifices. The likelihood function depends on the six parameters β_1, β_2, β_3, γ_1, γ_2 and γ_3 and can be maximized numerically.

Tolley *et al.* (1978) also chose a Weibull function to describe transition to the tumour state, but chose a Gompertz function for transition to death from other causes. More generally, Kalbfleisch *et al.* (1983) discuss likelihood estimation for an arbitrary parametric model without necessarily making the assumption of independent competing risks. In principle, however, the general formulation outlined earlier in this section, which does not require estimation of the distribution of time to death from causes other than tumour, seems simpler.

Limitations of multistage models and other modelling approaches

In an analysis of data from a mouse-skin painting experiment, it was assumed that the incidence rate followed a log-normal distribution with time (Day, 1967). Subsequent analysis by Peto *et al.* (1972) showed that Weibull distributions with a (κ, w)-pair common to all groups provided a significantly better fit to the data than did the log-normal.

Parish (1981) felt that it was unreasonable to expect animals to have an identical susceptibility to the effects of applied carcinogens, and suggested a model in which the parameter β had a gamma distribution. To gain an impression of the likely variation in susceptibility, she analysed data from the ageing experiment of Peto *et al.* (1975) referred to previously, looking at time to appearance of further tumours in animals according to how many tumours they already had. She concluded that the data were consistent with a 50-fold variation in susceptibility between the 5th and 95th percentile of the distribution. But even so, this variation was not large enough to make the distribution of time to tumour differ materially from a Weibull distribution, except where incidence rates were extremely high. The effect of susceptibility is to make the plot of log incidence against $\log(t - w)$ fall away from a straight line at high t values, and this may be why, in Figure 6.12, there is a discernible, slight drop-off with the 48 mg/week dose for the last four points plotted.

It has also been noted that in some circumstances the dose-response relationship is not of the form predicted by the multistage model. Davies *et al.* (1974), who tested response to seven dose levels of smoke condensate in a mouse-skin painting experiment, noted that there was a clear flattening off in response above doses of 180 mg/week. They suggested that high-dose levels were killing off a proportion of the cells at risk due to toxic effects, thus violating the first assumption of the multistage model that the number of cells at risk for each animal is the same for each group. In certain circumstances, it may be useful to modify the multistage model to allow for this possibility. Hulse *et al.* (1968) showed that the observed incidence of epidermal and dermal tumours in mice following superficial external β-irradiation may be accounted for by assuming that tumour incidence is proportional to the square of the dose and that potential tumour cells lose their reproductive integrity according to an exponentially decreasing relationship with dose. For example, the dose-dependent part of the hazard function may be of the form

$$(\beta_0 + \beta_1 d + \beta_2 d^2)\exp(-\alpha_1 d - \alpha_2 d^2).$$

Whittemore (1978) has reviewed a number of quantitative theories of carcinogenesis. She presented clear evidence of the inadequacy of theories not dependent on a multistage process, such as the single-stage theory of Iverson and Arley (1950) and the multicell theory of Fisher and Hollomon (1951), and considered a number of alternative versions of the multistage theory. She concluded that, although the multistage theory has a number of limitations (failure to distinguish between benign and malignant tumours, to consider the possibility of cell repair or the action of the host's immune system, to consider the differences in susceptibility or to consider that the sensitivity of target cells to transformation may not be constant), it nevertheless provides a flexible, broad and biologically plausible framework in which to examine the gross behaviour of tumour data.

Moolgavkar and Knudson (1981) propose a two-stage model which incorporates the growth and differentiation of normal target cells and intermediate cells (that is, cells in which the first stage has occurred). They demonstrate that experimental animal data and human epidemiological data are consistent with their two-stage model, noting that previous inferences that there are more than two stages in the development of cancer can be explained by differences in the growth kinetics of intermediate cells. By incorporating differentiation into their model, the authors are able to explain the age-incidence curves for some cancers (for example, certain childhood cancers), which cannot be explained easily in terms of a simple multistage model.

In an attempt to use the full information from an animal experiment (including time to death) for the estimation of 'safe doses', Hartley and Sielken (1977) model the hazard function as a product of a dose-dependent and a time-dependent term

$$\lambda(t, d) = g(d) \cdot h(t),$$

where the time-dependent term is chosen to be

$$h(t) = \sum_{r=1}^{R} rb_r t^{r-1}.$$

Proportional hazards models

For the analysis of rapidly lethal or observable tumours, we showed in Section 5.6 that the appropriate methods correspond to those given in Section 5.3 for survival analysis. The only difference is that one uses death due to tumour, or the appearance of an observable tumour, as the experimental endpoint, rather than death from any cause. It is useful to adopt the terminology 'failure' to denote such well-defined events as appearance of an observable tumour or death, and 'failure time' to denote the time to occurrence of such an event. Methods have been developed for the analysis of censored failure times which require no distributional assumptions (such as Weibull distributed time-to-tumour). The most widely used method is the proportional hazards model (Cox, 1972).

Under the proportional hazards model, the hazard function, that is, the age-specific failure rate for an animal with covariates $\mathbf{z} = (z_1, \ldots, z_p)'$, is

$$\lambda(t, \mathbf{z}) = \lambda_0(t) \cdot \exp(\boldsymbol{\theta}'\mathbf{z})$$

where $\lambda_0(t)$ is a completely unspecified hazard function, and $\boldsymbol{\theta}' = (\theta_1, \ldots, \theta_p)$ a vector of regression parameters, and

$$\boldsymbol{\theta}'\mathbf{z} = \sum_{u=1}^{p} \theta_u z_u.$$

For example, for a given animal, z_1 could be the administered dose level of a test compound, z_2 could be the initial body weight, z_3 could describe the row location and z_4 the column location of the animal's cage. Then the magnitude of the association between dose level of the compound and the failure, with adjustment for the remaining variables, can be measured by the estimate of the parameter θ_1 corresponding to z_1 in the proportional hazards model.

As in Section 5.3, suppose that failures are observed at K distinct times t_k, $k = 1, \ldots, K$. Let x_k denote the number of animals failing at t_k, and let s_k denote the sum of the covariate vectors z_{ik} corresponding to the animals failing at t_k, $i = 1, \ldots, x_k$. Then, if the number of ties (that is, $x_k > 1$) at each t_k is small, the parameter vector θ can be estimated by maximizing the approximate likelihood (Breslow, 1974):

$$L = \prod_{k=1}^{K} \frac{\exp(\theta's_k)}{\{\sum_{j \in R_k} \exp(\theta'z_j)\}^{x_k}},$$

where R_k denotes the set of indices corresponding to animals which survived to time t_k, and thus were at risk of failing at t_k. The approximate likelihood can be maximized using the Newton–Raphson method, and the covariance matrix for the resulting estimator $\hat{\theta}$ can be estimated by the negative of the inverse of the matrix of second partial derivatives of $\log(L)$ (Kalbfleisch & Prentice, 1980, Chapter 4; Miller, 1981b, Chapter 6). To illustrate analyses based on the proportional hazards model, consider the cigar-smoke condensate data (Section 4.2). The data can be examined for evidence of a dose-related increase in tumour rates by fitting the model $\lambda(t; z) = \exp(\theta z)\lambda_0(t)$, where $z = 0$ for animals in the low-dose group, $z = 19$ for animals in the middle-dose group and $z = 44$ for animals in the high-dose group. As noted above, the hazard function for the low-dose group, $\lambda_0(t)$, need not be specified. Maximizing the likelihood L leads to the estimate $\hat{\theta} = 0.0261 \pm 0.0060$, indicating that cancer risk increases with increasing dose.

Pairwise comparison of the middle-dose group to the low-dose group and of the high-dose group to the low-dose group can be accomplished by fitting a model which includes two covariates $z_1 = 1$ if the animal is from the middle-dose group and $= 0$ otherwise, $z_2 = 1$ if the animal is from the high-dose group and $= 0$ otherwise. The model

$$\lambda(t; z) = \exp(\theta_1 z_1 + \theta_2 z_2)\lambda_0(t)$$

provides estimates $\hat{\theta}_1$ and $\hat{\theta}_2$ such that $\exp(\hat{\theta}_1)$ and $\exp(\hat{\theta}_2)$ are estimates of the relative risk of the middle-dose group compared to the low-dose group, and of the high-dose group compared to the low-dose group, respectively. The results are summarized in Table 6.4.

Table 6.4 Estimates of relative risks using the proportional hazards model, for data on cigar-smoke condensate

	Dose d_i		
	0	19	44
$\exp(\hat{\theta}z_i)$	1.0	1.88	3.20
$\exp(\hat{\theta}_1 z_1 + \hat{\theta}_2 z_2)$	1.0	1.64	3.15

As in Section 5.6, there is good agreement between the estimates $\exp(\hat{\theta}_1 z_1 + \hat{\theta}_2 z_2)$ and the estimates based on the regression parameter θ.

If there are no ties (that is, $x_k = 1$ for all k), L can be derived as the marginal likelihood based on the distribution of ranks, but in the presence of ties, the expression for the marginal likelihood is more complicated than L (Kalbfleisch & Prentice, 1980, Chapter 4). Estimates of survival curves based on the proportional hazards model are available (Kalbfleisch & Prentice, 1980, pp. 84–87; Miller, 1981b, pp. 133–136), and estimates of percentiles can be obtained from the estimated survival curves. Confidence intervals for percentiles can be calculated by analogy to the methods based on the Kaplan–Meier survival curves (Slud et al., 1984), with substitution of the variance expression corresponding to the proportional hazards survival curves (Kalbfleisch & Prentice, 1980, pp. 116–117).

Regression models for tumour prevalence

The logistic regression model of Dinse and Lagakos (1983) for comparing treatment groups with respect to tumour prevalence was discussed in Section 5.5. Their regression model can be used in a more general setting when there are other covariates in addition to specific group membership or dose level. This more general regression context allows incorporation of several covariates and is computationally simple to analyse.

Formally, Dinse and Lagakos use the regression model

$$\psi(x, \mathbf{z}, t) = \frac{\mathrm{pr}(Y = 1 \mid X = x, \mathbf{Z} = \mathbf{z}, T = t)}{\mathrm{pr}(Y = 0 \mid X = x, \mathbf{Z} = \mathbf{z}, T = t)} = \exp\{\alpha x w(t) + \boldsymbol{\theta}'\mathbf{z} + \gamma(t)\}$$

to model the odds ratio of having $(Y = 1)$ or not having $(Y = 0)$ the tumour at death, with X a binary treatment indicator, $\mathbf{z}$ a p-vector of covariates and T survival time. The scalar α and the p-vector $\boldsymbol{\theta}$ are unknown parameters with $\gamma(t)$ and $w(t)$ prespecified functions of time. If treatment is assumed to have a constant effect on log-odds, $w(t)$ is set equal to 1 for all t. By setting $w(t) = 1 + f(t)\delta/\alpha$ where $f(t)$ is some function of time, such as t or $\log t$, one replaces $\alpha x w(t)$ by $\alpha x + \delta x f(t)$, and this allows a test of the hypothesis $\delta = 0$, that is, whether the proportional treatment odds relationship depends on time.

Integrated models for tumour prevalence and lethality

Another method, which differs from those described above, in that it takes into account the possibility of simultaneous study of more than one type of tumour, but is applicable only to experiments in which there are a number of scheduled sacrifices, has been described by Turnbull and Mitchell (1978) and by Mitchell and Turnbull (1979). For the purposes of their method, animals in one of R treatment groups $(r = 1, \ldots, R)$ dying in M time intervals $(m = 1, \ldots, M)$ are classified as being in one of K $(k = 1, \ldots, K)$ 'illness states'. These illness states are defined in terms of whether an animal has or does not have particular tumour types. Thus, dealing with three tumours of interest, there are $2^3 = 8$ illness states.

Their statistical model is defined in terms of prevalence p_{kmr}, the probability that an animal from group r, alive at the beginning of the mth interval, is in illness state k at that time, and lethality q_{kmr}, the conditional probability that an animal from group r dies during the mth interval, given that it was alive and in illness state k at the beginning of this interval. Data consist of w_{kmr}, the number of animals in group r sacrificed (withdrawn) in interval m and found in illness state k; d_{kmr}, the number of animals from group r dying in interval m and diagnosed with illness stage k, and s_{mr}, the number of animals from group r surviving through interval m. The distribution of such surviving animals among the illness states is not known, but can be characterized by s'_{kmr}, as these unobserved counts must be taken into account when fitting statistical models for the prevalence and for the lethality.

The numerator of the prevalence of an illness state is represented by $w_{kmr} + d_{kmr} + s'_{kmr}$, the term s'_{kmr} being estimated iteratively. Such values comprise data to which a log-linear model, formulated in terms of dependency of p_{kmr} on explanatory factors, such as treatment, time and the presence of tumours of each type can be fitted.

Similarly, a statistical model is fitted for lethality q_{kmr}. Since the lethality is a conditional probability, both numerator, d_{kmr}, and denominator, $d_{kmr} + s'_{kmr}$, have to be specified. The model used is a logistic one in which the dependency of q_{kmr} on a similar set of explanatory factors is studied.

The crucial problem of not knowing the s'_{kmr} is dealt with as follows. Firstly, the distribution of the s_{mr} survivors to the illness states is assumed to be as in the distribution observed in the animals which died or were killed, so that

$$s'_{kmr} = s_{mr}(d_{kmr} + w_{kmr}) \bigg/ \sum_{k=1}^{K} (d_{kmr} + w_{kmr}).$$

Using these s'_{kmr}, prevalence and lethality models are then fitted, which give rise to estimates $\hat{p}_{kmr}$ and $\hat{q}_{kmr}$. These estimates are then used to re-assess the distribution of the s_{mr} survivors to the illness states by the formula

$$s''_{kmr} = s_{mr}\hat{p}_{kmr}(1 - \hat{q}_{kmr})/(1 - h_{mr}),$$

where $h_{mr} = \sum_{k=1}^{K}\hat{p}_{kmr}\hat{q}_{kmr}$ is the unconditional probability of dying in group r in interval m. s'_{kmr} can be replaced by s''_{kmr} and the same models fitted to the slightly modified data. This process can then be iterated until the distribution of the survivors into the illness state no longer changes.

It should be noted that the analysis makes the (technical) assumption that illness state changes are made only at the beginning of each of the M time intervals and that sacrifices occur only immediately after the illness state changes, in order for the prevalences and the lethalities to be defined. It is thus convenient to choose partitions of the time axis such that each interval covers one scheduled sacrifice.

A detailed application of this approach is given by Wahrendorf (1983). A very interesting feature of this method is that it allows simultaneous analysis of different tumour types. This opens interesting possibilities for assessing associations between different tumours. Berlin *et al.* (1979) also consider a general Markov model for multiple tumour types and discuss the question of identifiability.

6.4 Summary

The number of animals developing tumours during the course of a conventional two-to-three-year rodent carcinogen bioassay will depend on the dose level to which they are exposed. The overall shape of the dose-response curve can, however, vary widely, depending on the particular agent being evaluated. Although most dose-response relationships generally increase with dose, this increase may be either linear or distinctly nonlinear. In the latter case, dose-response curves which increase rapidly beyond a certain dose range may be noted, as with nasal tumours induced by exposure to formaldehyde. Conversely, a levelling off of the rate of response may be observed at high doses, as with liver tumours resulting from exposure to vinyl chloride. Combinations of these different shapes are also possible, as with the S-shaped dose-response seen for aflatoxin-induced liver tumours.

A variety of different mathematical dose-response models may be used to provide a parsimonious description of the observed dose-response relationship. Simple tolerance distribution models, although often sufficiently flexible to provide a good fit to dose-response data, are somewhat naive in terms of their underlying biological basis as a possible mechanism for carcinogenesis. Stochastic models based on the notion that carcinogenesis results from the random occurrence of one or more fundamental biological events are more appealing, but are necessarily based on strong but uncertain assumptions. Foremost among such mechanistic models is the Armitage–Doll multi-stage model, based on the assumption that a cell progresses through a number of distinct stages before becoming cancerous, the transition intensity function for each stage being a linear function of dose.

Since many compounds may require metabolic activation before being converted into their active form, consideration may be given to pharmacokinetic models for this process. This is particularly important when certain steps such as absorption, elimination, activation or detoxification are saturable. Even if the response rate is directly proportional to the dose level of the activated complex reaching the target tissue, such saturation effects may account for nonlinearity in the dose-response relationship when expressed as a function of the administered rather than of the delivered dose.

Given a suitable model for the dose-response relationship, estimates of certain quantiles of the curve may be of interest. Because human exposure to most environmental carcinogens is low, there has been considerable interest in obtaining estimates of risk in the low-dose region based on the downward extrapolation of results obtained at higher doses. Unfortunately, this is subject to considerable model uncertainty, and different models, all equally consonant with the observed data, give widely different projections upon extrapolation of low doses. Because of this, a linear extrapolation to low doses is often advocated as the most prudent approach. This will be particularly appropriate in cases where the background response rate can be considered to arise from an at least partially dose-wise additive model.

This low-dose model dependency may be circumvented by restricting attention to those quantiles lying well within the observable response range. Historically, quantiles such as the TD_{50} (the dose estimated to induce tumours in 50% of exposed animals) have been used as the basis for various measures of carcinogenic potency. The TD_{50} itself has recently been used by Gold *et al.* (1984) to demonstrate variations in carcinogenic potency spanning eight orders of magnitude.

Time-to-tumour models attempt to describe the carcinogenic process in detail and make use of information on individual tumour occurrence and survival times. For observable tumours, parametric models addressing the time to first occurrence of this tumour have been developed on biological principles. The parameters of these models allow for inferences to be drawn on both the time and dose dependency of tumour incidence. For occult tumours the situation is more complicated: tumour incidence is not directly observable and its estimation depends on the experimental design, particularly the use of a series of interim sacrifices. However, within this framework, estimates for the relevant functions characterizing the carcinogenic response are derived, if identifiable. This provides the basis not only for unbiased statistical inference but also for clear biological conclusions. Time-to-tumour models make use of as much information as possible; they are in turn based on some assumptions, but they provide the most thorough description of the observed response in a long-term animal experiment.

LIST OF ESSENTIAL SYMBOLS – CHAPTER 6 (in order of appearance)

$P(d)$	probability of treatment-induced tumour at dose d
$G(t)$	tolerance distribution
$P^*(d)$	probability of spontaneous or treatment-induced tumour at dose d
π_0	background response rate (probability of spontaneous tumour)
$D_1(t)$	administered dose at time t
$D_2(t)$	activated dose at time t
D_2^*	effective dose under steady state conditions
θ	vector of parameters $\theta_1, \ldots, \theta_p$
$P^*(d, \theta)$	probability of response (tumour) at dose d dependent on parameter θ
d_i	dose levels ($d_0 = 0, d_1 < \cdots < d_I$)
x_i	number of animals with tumour at dose d_i
n_i	number of animals at dose d_i
$L(\theta)$	likelihood of observed outcome under any dose-response model $P^*(d; \theta)$
$\hat{\theta}$	maximum likelihood estimate of θ
0	vector of zeros
V	variance–covariance matrix of $\sqrt{n}\,(\hat{\theta}^1 - \theta)$
v^{rs}	(r, s)th element of the inverse of V
$\Pi(d)$	added risk over background at dose d
d_q	dose corresponding to the added risk, q, over background
C_q	measure of carcinogenic potency based on d_q
N	disease-free state of animal
T	animal with tumour
D_{NT}	death from a cause unrelated to the tumour of interest
D_T	death due to tumour of interest

X	time to onset of tumour (random variable)
Y	time to death due to tumour (random variable)
Z	time to death from an unrelated cause (random variable)
$f(t)$	density function
$\lambda(t)$	hazard function
$\Lambda(t)$	cumulative hazard function
$S(t)$	survival function

(N.B. A subscript X or Y to the quantities $f(t)$, $\lambda(t)$, $\Lambda(t)$ and $S(t)$ indicates the corresponding functions for the random variable X or Y)

t_k	time of observation of an event such as death or the occurrence of tumour $(k = 1, \ldots, K)$
a_k	number of animals with appearance of a visible tumour at time t_k
b_k	number of animals with death caused by tumour of interest (fatal context) at time t_k
c_k	number of animals with death from unrelated cause, tumour of interest present (incidental context), at time t_k
d_k	number of animals dying without tumour of interest at time t_k
L_1	likelihood in the case of visible tumours
L_2	likelihood in the case of occult tumours all observed in a fatal context
L_3	likelihood in the case of occult tumours all observed in an incidental context
L_4	likelihood in the case of occult tumours observed in fatal or incidental context

(N.B. LL_i $(i = 1, 2, 3, 4)$ denotes the logarithm of the likelihoods L_i $(i = 1, 2, 3, 4)$)

κ	number of stages or transformations required before tumour occurs
δ_j	kinetic rate constants for the jth stage of multistage model $(j = 1, \ldots, \kappa)$
w_j	constant time taken by the jth transformation of the multistage model $(j = 1, \ldots, \kappa)$
β_i	Weibull shape parameter for group i $(i = 1, \ldots, I)$
LL	log-likelihood for an experiment with I groups under the Weibull model with known w and κ
$\mathbf{z}$	vector of covariates observed for each animal
$\lambda(t; \mathbf{z})$	hazard function of animal with covariates $\mathbf{z}$
$\lambda_0(t)$	baseline hazard function

7. SPECIAL TOPICS

CHAPTER 7

SPECIAL TOPICS

7.1 Introduction

In the previous chapters we considered the broad aspects of the design and analysis of long-term animal experiments. There are, however, many special aspects that have not been considered in the previous chapters, but that merit mention, as they occur frequently in practice. These aspects are all concerned, in one way or another, with multiplicity; that is, in design or analysis, one factor or one variable is added or allowed to have multiple levels.

In Section 7.2, we shall deal with the problem that arises from the study of more than one tumour of interest in a long-term animal experiment, and, thus, the statistical inference has to be carried out on data from a variety of sites. In some experimental systems, the multiplicity of tumours at one site is viewed as a relevant biological endpoint, and the appropriate statistical methods are discussed in Section 7.3. A tumour response may be graded according to a fixed number of clinical or histological categories, and the methods applicable to these situations are outlined in Section 7.4. If a further dimension is added to the design of an experiment, either by stratification or by adding another exposure factor, it must be considered in the analysis, and methods for doing so are outlined in Section 7.5. In some experiments, the information about the common litter membership of the test animals may be kept in order to influence the design of the experiment. Statistical methods making use of this information are discussed in Section 7.6. Several tumours can occur simultaneously in one animal; in Section 7.7, we outline methods for assessing the association among tumour types in appropriate experiments. Finally, Section 7.8 deals with the incorporation of historical information on tumour incidence in untreated control animals into the statistical analysis.

7.2 Statistical inference at multiple sites

The false-positive and false-negative rates are of great importance in any screening procedure. In a carcinogenesis screening test, the false-positive rate is the percentage of noncarcinogenic compounds which are incorrectly classified as carcinogens, and the false-negative rate is the percentage of carcinogenic compounds which are incorrectly classified as noncarcinogens. In most animal carcinogenesis experiments, it is not possible to predict *a priori* potential target organs at which carcinogenic effects are likely to occur. Thus, although the effect of a carcinogenic agent is likely to be concentrated in one or a few target organs, all organs which are examined histopathologically must be evaluated for evidence of carcinogenesis. Because of the multiple comparisons involved in the statistical evaluation of tumour incidence data from several organ and tissue sites, there is a danger of inflating the false-positive error rate. In particular, a simplistic decision rule that routinely labels a chemical as a carcinogen whenever a single tumour increase is significant at the 5% level for any exposed group at any of the organs examined can result in a false-positive rate considerably greater than 5% (Fears et al., 1977; Salsburg, 1977; Fears & Tarone, 1977; Haseman, 1977; Elashoff et al., 1979). Some authors have argued that the inflated false-positive rate associated with such a naive decision rule invalidates the use of animal experiments in screening chemicals for carcinogenesis (Salsburg, 1977). Others have noted that such a decision rule is not in fact used in practice, and that rules which attempt to model the actual decision process indicate that false-positive rates are close to the nominal level (Fears et al., 1977; Fears & Tarone, 1977; Haseman, 1977; Gart et al., 1979; Haseman, 1983b).

A major problem in trying to estimate the error rates of a carcinogenesis screening test, or to recommend explicit adjustments for the multiple comparisons involved, is the difficulty in modelling the interaction that takes place between the statistician and scientists of other disciplines included in the decision-making process. The evaluation of the carcinogenic potential of a test compound is not strictly a statistical decision. It is impossible to incorporate the totality of knowledge and experience of the pathologists, toxicologists, pharmacologists and other scientists involved in the decision-making process into a simple statistical model. Nevertheless, investigations based on the comparison of unadjusted tumour rates using the Fisher–Irwin exact test have led to statistical devices that can be used to keep false-positive rates under control. The most important finding of these investigations is that organs with low spontaneous tumour rates can be ignored effectively in the calculation of false-positive rates (Fears et al., 1977; Gart et al., 1979; Haseman, 1983b). In particular, for a given experimental design and nominal significance level, one can compute the minimum number of animals, in the combined control and exposed groups, which must be found with a given tumour in order for a significant result to be obtained using the Fisher–Irwin exact test. Accordingly, only those organs with spontaneous tumour rates for which this minimum number of tumours is likely to be obtained need be considered in determining an adjustment for multiple comparisons (Gart et al., 1979). An alternative, but related, approach to control for multiple comparisons is to test for tumour increases using one nominal significance level, say $\alpha_1 = 0.05$, for organs with low spontaneous tumour rates (for example, tumour rates less than 2% for experiments with 50 animals per group), and a second, smaller nominal significance level, $\alpha_2 < 0.05$, for all other organs. The actual value of α_2 leading to a false-positive rate of 5% can be calculated for each species/strain/sex combination for which good estimates of spontaneous tumour rates exist. Various modifications of this approach are possible, for example, using a different nominal level for each value of the spontaneous tumour rate, the nominal level decreasing with increasing tumour rate. Of course, if there is evidence a priori that a test compound is likely to produce a carcinogenic effect at a particular organ, then the nominal significance level for the suspected organ should not be reduced, regardless of the magnitude of the associated spontaneous tumour rate.

In order to avoid the multiple comparisons problem, Brown and Fears (1981) proposed a method of calculating a single overall significance level for a carcinogenesis experiment. Suppose that, in an experiment with one exposed group and a concurrent control group, T organs are examined for the presence of a tumour in each animal. Then, each animal may have tumours discovered in one of 2^T possible combinations of organ sites, ranging from 'no tumours found' to 'tumours found in all T organs'. For each of the T organs, a Fisher–Irwin exact test can be performed. For a fixed significance level, α, they provide a method for calculating the exact permutational probability of at least one significant Fisher–Irwin test, conditional on the 2^T marginal totals (each marginal total is the number of animals with tumours only in the organs represented by one of the 2^T possible combinations). An overall significance level can be calculated by applying the method with α set equal to the smallest of the T p-values observed in the individual Fisher–Irwin exact tests. Unlike previously discussed methods, the method of Brown and Fears requires no prior knowledge of the spontaneous tumour rates.

Meng (1985) has proposed a Bayesian approach to the multiple comparisons problem, incorporating historical data on spontaneous tumour rates in a manner suggested by Dempster *et al.* (1983). The method proposed by Meng has the disadvantage that an increased tumour rate at a single organ can be diluted by a general decrease in tumour rates at other organs (such general decreases have been observed due to the reduced food consumption in exposed animals). However, the development of related methods warrants further investigation.

7.3 Multiplicity of tumours

The methods described in Chapter 5 concentrate on the presence or absence of one or more tumours in an animal. This reflects the fact that the fundamental measure of carcinogenic effect is usually taken as the total number of *animals* which develop a tumour of a given type rather than the total number of such *tumours* (Peto *et al.*, 1980). The main reasons for this are, firstly, that multiple tumours in an animal are not independent events (a few animals often get a large number of tumours) and, secondly, that for tumours not observable until death it is impossible to determine whether treatment has caused tumours or has merely affected their progression. In theory, as noted by Peto *et al.* (1980), a chemical which inhibits metastatic spread of localized tumours might allow animals with tumours to live longer and have time to develop more tumours. Furthermore, it is the individual animal that is randomized among the dose groups, and thus the animal should be treated as the experimental unit.

In some cases, however, experimental systems have been specifically developed to quantify response in terms of multiplicity, and it is useful to have methods available which take into account the number of tumours at a given site. The most widely used system of this type involves the mouse skin, where topical application of carcinogens can produce a sequence of multiple lesions – usually papillomas. Continuous surveillance of the animals is necessary to observe the course of the lesions accurately.

This involves observing the times of first occurrence of the lesions and the times when some disappear due to systemic regression, scratching, biting or other external reasons.

Another experimental model developed by Shimkin (Shimkin & Stoner, 1975) measures the development of lung adenomas in mice in a relatively short period of time. Animals are killed after seven or eight months and the number of lung adenomas are counted as a quantitative endpoint. In a long series of experiments with urethane, this same model was used to try to elucidate the mechanism of action of urethane carcinogenesis; in direct screening assays with this model, urethane is usually considered as a positive control.

The induction of multiple mammary tumours in female Sprague Dawley rats, mainly by 7,12-dimethylbenz[a]anthracene, is an animal model which has been developed to study the possible inhibitory effect of other chemicals such as, for example, vitamins, on carcinogenesis. Multiplicity of tumours in this model is considered a quantification of the response. This rat mammary model was developed as a quick model for direct screening of compounds, but it has lost favour for this purpose due to its limited specificity.

In general, none of these special animal models are considered to provide conclusive evidence on their own when used for screening the carcinogenicity of chemicals. They play a more useful role in the study of the mechanisms of carcinogenesis. Nevertheless, there is an interesting challenge in using the appropriate methodology for analysing such studies.

It is necessary to distinguish between the situation in which the number of tumours is counted at a fixed point in time in each experimental group and that in which the time of development of each individual tumour is accurately recorded. In the first case, the number of tumours seen in an individual animal represents the basic information. Let x_{li} denote this number for the lth animal in the ith experimental group ($l = 1, \ldots, n_i$; $i = 0, \ldots, I$). Analysis of variance methods can be applied to such data. Both parametric and nonparametric methods are available to test the null hypothesis that there is no difference between the experimental groups against either the unstructured alternative that the responses are different between groups, or the ordered alternative that the responses are increasing with increasing dose level.

In Section 8.3, detailed methods are given for the analysis of concomitant information by parametric or nonparametric one-way analysis of variance. These methods can be applied directly by treating the tumour counts x_{li} ($l = 1, \ldots, n_i$; $i = 0, \ldots, I$) as the basic observation per animal. In a parametric analysis of variance, the x_{li}'s are used directly for the calculation of the test statistics, whereas in the nonparametric methods they are converted into ranks. In the first case, one may also apply transformations, for example the square root or logarithm, to achieve a better fulfilment of the underlying assumptions (equal variances, normal distribution of observational errors).

One parametric approach has been proposed by Drinkwater and Klotz (1981). They suggest that the number of tumours per animal has a Poisson distribution, that is, the probability, $f(t)$, that an animal bears t tumours is

$$f(t) = e^{-\lambda}\lambda^t/t! \qquad (t = 0, 1, 2, \ldots).$$

Furthermore, in order to account for the empirically observed variation, they suggest that the parameter λ, which is the mean number of tumours per animal, is subject to further random variation modelled by a gamma distribution. This leads to a so-called negative binomial distribution which is frequently applied in the analysis of count data (Anscombe, 1949, 1950; Bliss, 1953). Drinkwater and Klotz (1981) outline the calculation of a likelihood ratio test statistic to compare the tumour counts in two experimental groups under this parametric model. Their comparison of this method with the *t*-test (parametric analysis of variance), the Wilcoxon test (nonparametric analysis of variance) and the chi-square test from a 2×2 table contrasting tumour incidence in the two groups is not fully conclusive as it is based on simulated data derived exactly from the model of a negative binomial distribution. However, they do provide some empirical support for this model. In the absence of any firm knowledge about a parametric model for the variation of the tumour counts, we recommend a nonparametric analysis of such tumour counts at a fixed point in time.

When the time of appearance of the multiple tumours is recorded for each animal, the methods developed by Gail *et al.* (1980) can be used. They consider that, in each animal, tumours are observed to appear at times $T_1 < T_2 < \cdots < T_K$. The notation of capital letters indicates that we introduce their approach in terms of the observable random variables. In addition, for any animal there is a censoring time C which is assumed to be independent of the sequence $T_1, T_2, \ldots, T_K$, K being the largest integer such that $T_K < C$. The *j*th gap is defined as $Z_j = T_j - T_{j-1}$, with $T_0 = 0$ for convenience. If the probability distribution of the *j*th gap depends only on j and on t_{j-1}, the value of T_{j-1}, but not on the earlier times $t_1, \ldots, t_{j-2}$, the sequence T_1, $T_2, \ldots$ is called a Markov sequence. The hazard function of Z_j is denoted by $h(z \mid j, t_{j-1})$ and is used as the basic element in developing inferential strategies. For this purpose, the authors consider that $h(z \mid j, t_{j-1})$ has a known parametric form or that $h(z \mid j, t_{j-1})$ is independent of t_{j-1}, in which case a so-called semi-Markov model results. In both situations the effect of the different treatment groups on the occurrence of tumours is modelled similarly to the proportional hazards model (Cox, 1972). To keep the notation simple, we consider two treatment groups, 1 and 2. Conditional on t_{j-1}, it is assumed that the gap Z_j has hazard

$$\exp(\alpha_j) h(z \mid j, t_{j-1}) \quad \text{or} \quad h(z \mid j, t_{j-1})$$

according to whether the animal has been given treatment 1 or 2, respectively. Gail *et al.* (1980) discuss methods of estimating the parameters α_j, of testing their homogeneity, and of testing that the common value of α is zero. As mentioned above, various specializations of $h(z \mid j, t_{j-1})$ are used.

The models used in this approach can be derived from the so-called *m*-site model, which is often used in mathematical theories of carcinogenesis (Whittemore & Keller, 1978). The original paper should be consulted in detail when applying these methods. An alternative strategy is to apply the proportional hazards model (Cox, 1972) (see Section 6.3) with its feature of time-dependent covariates or strata (Kalbfleisch & Prentice, 1980) to adjust for the number of tumours already developed while comparing the hazards of developing the next tumour (Scribner *et al.*, 1983).

7.4 Graded responses

As the principal interest in carcinogenicity experiments is the presence or absence in an animal of a tumour of a given type, we have concentrated on techniques to analyse response as a 0–1 variable. Section 7.3 dealt with the multiplicity of tumours where counts of tumours occurring at a given site represent a quantification of the response. Another quantitative measure of the carcinogenic response is the grade of the lesion according to some pathological criteria. These may differ from site to site and between schools of experimental pathology, but it is possible to grade any lesion on a scale, such as: $0 =$ absent, $1 =$ minimal, $2 =$ slight, $3 =$ moderate, $4 =$ severe, $5 =$ very severe; or, $0 =$ absent, $1 =$ benign, $2 =$ malignant.

In the case of grading, a fixed, limited scale is applied to all animals, whereas for multiple tumour counting (Section 7.3) there is, at least in principle, no limitation on the number of tumours which could be observed.

Snedecor and Cochran (1980, pp. 146–148) have suggested that comparison of graded data from two groups may be carried out using Fisher's randomization test with small numbers and a continuity-corrected t-test with larger numbers. In the latter case, one could alternatively use the nonparametric techniques discussed in Section 8.3. Application of ranking procedures to graded responses can lead to a large number of ties, but it has been stated that this may not be crucial (Conover, 1980, p. 232).

To illustrate the suggestion of Snedecor and Cochran, consider the simple, fictitious experimental outcome (Table 7.1) of a control group of sample size six and a dose group of sample size four, graded on a three-point scale, 0, 1, 2.

Table 7.1 Example to illustrate Fisher's randomization test

n_i	Control group (6)	Dose group (4)
(x_{ij})	(0, 0, 0, 0, 1, 1)	(0, 1, 1, 2)
$x_{i.}$	2	4
$\bar{x}_{i.}$	$\frac{2}{6} = \frac{1}{3}$	$\frac{4}{4} = 1$

The randomization test is based on the fact that there are $(n_0 + n_1)!/(n_0!n_1!)$ possible divisions of the $n_0 + n_1$ animals into groups of n_0 and n_1. We also wish to find the number of such possible outcomes for which $\bar{x}_{1.} - \bar{x}_{0.}$ matches or exceeds the observed value. Equivalently, this is the number of outcomes for which $x_{1.}$ matches or exceeds the observed value, for example, 4. In this particular case, the number of outcomes is $10!/(4!6!) = 210$, the observed $\bar{x}_{1.} - \bar{x}_{0.} = \frac{2}{3}$ and $x_{1.} = 4$. Consider the numbers of ways in which $x_{1.} \geq 4$. These are listed in Table 7.2.

Thus, there are in total $30 + 1 + 4 = 35$ combinations with $x_1 \geq 4$, and the exact one-tailed p-value is $p = 35/210 = \frac{1}{6}$. The corresponding t-test with $n_0 + n_1 - 2$ degrees of freedom is

$$t_c = (\bar{x}_{1.} - \bar{x}_{0.} - c)/\{s_p\sqrt{(1/n_0 + 1/n_1)}\},$$

where

$$s_p^2 = \left\{\sum_{j=1}^{n_1}(x_{1j} - \bar{x}_{1.})^2 + \sum_{j=1}^{n_0}(x_{0j} - \bar{x}_{0.})^2\right\}\bigg/(n_0 + n_1 - 2)$$

Table 7.2 List of extreme outcomes for computation of randomization test

	(x_{1i})	Numbers of combinations
$x_{1.} = 4$	(0, 1, 1, 2)	$\binom{5}{1}\binom{4}{2}\binom{1}{1} = 30$
	(1, 1, 1, 1)	$\binom{5}{0}\binom{4}{4}\binom{1}{0} = 1$
$x_{1.} = 5$	(1, 1, 1, 2)	$\binom{5}{0}\binom{4}{3}\binom{1}{1} = 4$

and c is a continuity correction. The value of c is one-half of the absolute value of the difference between the observed $\bar{x}_{1.} - \bar{x}_{0.}$, and the next highest value of this statistic among all the possible randomizations. Thus, for this example, $\bar{x}_{1.} - \bar{x}_{0.} = \frac{4}{4} - \frac{2}{6} = \frac{2}{3}$, and the next highest value is $\frac{3}{4} - \frac{3}{6} = \frac{1}{4}$. Thus $c = (\frac{2}{3} - \frac{1}{4})/2 = \frac{5}{24}$. We also find that $s_p^2 = \frac{5}{12}$. These give the approximate t-value with eight degrees of freedom:

$$t_c = \frac{\frac{2}{3} - \frac{5}{24}}{\sqrt{\{(\frac{5}{12})(\frac{1}{4} + \frac{1}{6})\}}} = \frac{11}{10} = 1.1.$$

This approximation yields $p = 0.1517$, only slightly less than the exact result of $\frac{1}{6}$.

It should be noted that Snedecor and Cochran recommend that with such small samples with very little overlap the exact p is easily calculated and should be reported. They note that, with more extreme results, t_c may yield too small a p-value. Consider the most extreme possible outcome in our example, that is, $x_{1.} = 5$ or $\bar{x}_{1.} - \bar{x}_{0.} = \frac{5}{4} - \frac{1}{6} = \frac{13}{12}$. The exact one-tailed $p = 4/210 = 0.0190$. In this case,

$$t_c = \frac{\frac{13}{12} - \frac{5}{24}}{\sqrt{\{(\frac{10}{97})(\frac{1}{4} + \frac{1}{6})\}}} = 3.047,$$

which, with 8 degrees of freedom, gives $p = 0.0079$, less than one-half the exact p-value. Randomization tests for trend and associated approximate t-tests may also be performed when more than two groups are to be compared.

The particular nature of a graded response may also be taken into account by using methods which have been developed for the analysis of ordinal data (McCullagh, 1980). Let the response be graded in G categories, $g = 1, \ldots, G$, and let there be $I + 1$ experimental groups, represented by dose levels $d_0, d_1, \ldots, d_I$. As above, let n_{gi} be the number of lesions in group i ($i = 0, 1, \ldots, I$) at grade g ($g = 1, \ldots G$). From these data, one can estimate γ_{gi}, the probability of a lesion in group i being graded at or below level g. Note that $\gamma_{Gi} = 1$, so that there are only $G - 1$ essential estimates γ_{gi} ($g = 1, \ldots, G - 1$) for each group.

McCullagh (1980) proposes finding a suitable transformation of γ_{gi} and investigating how the transformed values depend on the dose levels d_i. Two particular models have been proposed for this purpose. One model is

(i)   the proportional odds model:

$$\log[\gamma_{gi}/(1 - \gamma_{gi})] = \theta_g - \beta d_i \qquad (1 \le g < G).$$

This means that for any two dose levels i_1 and i_2 the odds ratio

$$[\gamma_{gi_1}/(1 - \gamma_{gi_1})]/[\gamma_{gi_2}/(1 - \gamma_{gi_2})] = \exp \beta(d_{i_2} - d_{i_1})$$

is independent of the grade g and depends on the difference between the dose levels only. The parameters θ_g, which are nuisance parameters, and β, which is the essential parameter relating the graded response to the dose levels, can be estimated by maximum likelihood methods. Another model is
(ii)   the proportional hazards model

$$\log[-\log(1 - \gamma_{gi})] = \theta_g - \beta d_i \qquad (1 \le g < G).$$

This model is related to the proportional hazards model formulated by Cox (1972) for the analysis of survival data. Again, maximum likelihood methods allow estimation of the parameters θ_g and β.

In general, any other monotone increasing function mapping the unit interval $(0, 1)$ onto $(-\infty, \infty)$ can be used as a 'link'-function $l(\gamma)$ to postulate a model

$$l(\gamma_{gi}) = \theta_g - \beta d_i \qquad (1 \le g < G).$$

The two examples given above, however, have the advantage of a straightforward interpretation of the parameters. This regression model can be generalized to allow for any set of covariates, not only one dose variable. The number of parameters will increase accordingly.

McCullagh has also pointed out that there is a direct theoretical relationship between his regression models for ordinal data and the nonparametric test discussed above, the latter, however, lacking simple descriptive parameters.

Graded response data can also be analysed in a stratified way, for example, when a tumour observed in an incidental context is graded. Generally, the grading is more severe in later time periods, and fewer animals from the higher-dose groups survive into the later intervals, due to toxicity. In such a case, the analysis could be stratified by time intervals, the θ_i's varying between time intervals, but the essential parameter β being the same for all.

7.5   Multifactorial designs: combining results

In the previous chapters, attention has centred on the design and analysis of experiments that test only one treatment of interest, often at different dose levels. Such one-factor experiments are commonly used to screen different exposures for their potential carcinogenicity. However, studies in which the experimental groups form a multiple-factor design are not uncommon. Sometimes, such designs are necessitated by practical reasons, so that, for example, the eight groups might form combinations of the main exposure of interest at four levels (control, low dose, middle dose, high dose) and two batches of animals, as it is impossible to obtain the number of animals required from one batch. On other occasions, there may be one main treatment of interest, but one may wish to study simultaneously the effect of different methods of administering the treatment, for example, cigarette-smoke condensate at three dose levels dissolved in two alternative solvents. More interestingly, one may wish to study the effects of joint exposure to combinations of two or more carcinogens. It can be argued that such studies can often be more realistic than single carcinogen studies, since humans are frequently exposed to a variety of carcinogens simultaneously or in sequence. Often such studies are of value in investigating mechanisms of action; they also have direct public health implication, and they are often carried out where knowledge has already been accumulated about the dose-response of individual exposures.

In Chapter 3, the design of multifactorial experiments was briefly mentioned and some formal concepts outlined. For the purposes of this section, we limit ourselves to the study of two exposures, A and B, applied at all combinations of dose levels $a_0(= 0)$, $a_1, \ldots, a_i, \ldots, a_I$ of A and $b_0(= 0)$, $b_1, \ldots {}_\Delta b_j, \ldots, b_J$ of B so that there are $(I + 1)(J + 1)$ experimental groups. Such a design leads to two distinct questions:

(1) Given equivalent exposure to B, is the risk of tumour significantly related to exposure A?

(2) Is the joint effect of A and B different from what one would expect from the effect of A or B alone?

The first question is essentially aimed at avoiding bias due to the effect of B in assessing the effect of A, and of making a combined inference about A over the different levels of B. Seen in this light, A is the main exposure of interest, B being a secondary 'nuisance' variable that has to be standardized for.

There are two major techniques for answering this first question. Throughout Chapter 5, we have extensively described methods in which observed and expected values (as well as other statistics necessary for calculating significance levels of the observed/expected differences) from different time periods can be combined by accumulation. As long as the group structure remains the same in each stratum over which accumulation occurs, this method of combining can be used in an exactly analogous manner to combine results in dimensions other than time. Thus, in our example, we treat the data as consisting of $J + 1$ subexperiments ('strata') defined by the levels of B. Each subexperiment has the same group structure ($I + 1$ levels of A), and a combined result can be obtained in a straightforward manner. The same process, of course, can be used to make overall inferences for exposure B, adjusted for the effects of exposure A.

The second major technique for answering the first type of question would be applied where the response variable of interest can be related to the effects of the exposures A and B by a regression equation. For this purpose, we introduce the general notation that u_{ij} is the response in those animals exposed to level i of exposure A and level j of exposure B. For specific applications, the effect measure u_{ij} has to be defined very carefully, and this will have strong implications on the interpretation of the results. However, we discuss first the general concepts of absence or presence of interactions by denoting further μ to be an overall mean response, t_i a deviation from the mean due to the ith level of A, and c_j a deviation from the mean due to the jth level of B (t_0 and c_0 are assumed to be zero to avoid overparameterization).

A test of the effect of A adjusted for the effect of B can be achieved by comparing the fit of the models

$$E(u_{ij}) = \mu + t_i + c_j$$

and

$$E(u_{ij}) = \mu + c_j,$$

where $E(u)$ denotes the expectation (i.e., mean) of u. This is, in general, a more appropriate test for the effect of A than the comparison of the two simpler models both ignoring c_j. The test recommended, on I degrees of freedom, is a test of overall variation in response with level of exposure A. It is of course possible to test for a linear effect of A by replacing t_i in the above formulation by a term γd_i, where γ is a parameter to be estimated and d_i is the dose applied at level i, although the full set of s parameters, representing effects of the confounding variable B, should be retained.

Implicit in both techniques for answering the first type of question is the no-interaction assumption, that is, that the effect of A does not vary significantly according to level of B. An overall conclusion that exposure A slightly increases tumour risk might be misleading, for example, if it considerably increased risk at high doses of B, while reducing risk somewhat at low doses. Statistical tests for interaction of the effects of the stratifying variable with those of the main variable of interest, when analysing stratified contingency tables, are given by Breslow and Day (1980). With the regression equation, a test of no interaction can be achieved by comparing the fit of the model

$$E(u_{ij}) = \mu + t_i + c_j$$

with that of the model

$$E(u_{ij}) = \mu + x_{ij},$$

where x_{ij} represent effects of each combination of treatments (x_{00} is assumed to be zero to avoid overparameterization).

In the second question, the interest is in the joint effect of both exposures. This, of course, is related to the test of no interaction, as lack of interaction implies in a sense that the joint effect of A and B does not differ from that of A or B, alone, since the data are well described by the model

$$E(u_{ij}) = \mu + t_i + c_j.$$

It is important to repeat that the particular type of model depends strongly on the scale on which the effects u_{ij} are measured. To consider this further let us for the moment take the response variable in experimental group (i, j) to be the probability of developing a tumour p_{ij}, which is estimated by the proportion of tumour-bearing animals. p_{i0} ($i = 0, 1, \ldots I$) and p_{0j} ($j = 0, 1, \ldots J$) then denote the probabilities of the dose-response patterns for the individual exposures.

There are two basic possibilities for modelling the response probability p_{ij} in the combination groups exposed to both A and B.

(i) In the first, *the additive model,* we consider the absolute increase over the background response probability, measured in the untreated control group, as the quantity which describes the effect of exposure, and assume that, in a group treated with both exposures, this effect should be the sum of the effects of both the respective individual effects

$$p_{ij} - p_{00} = (p_{i0} - p_{00}) + (p_{0j} - p_{00})$$

or

$$p_{ij} = p_{i0} + p_{0j} - p_{00}$$

for i and $j > 1$, p_{ij} being taken as 1 if the right-hand side of the second equation exceeds 1.

(ii) In the second, *the multiplicative model,* the effect of an exposure is measured by the proportional increase over the background response, so that in a combination group the effect should be equal to the product of the effects of individual exposures

$$p_{ij}/p_{00} = (p_{i0}/p_{00})(p_{0j}/p_{00})$$

for i and $j > 1$. This is also equivalent to

$$p_{ij}/p_{0j} = p_{ik}/p_{0k}$$

for $j, k = 0, 1 \ldots, J$ $(j \neq k)$, which means that the effects attributed to exposure A indicated by the subscript i are the same at any level j or k of exposure B.

The two models introduced above represent simple statistical models based on the choice of different effect measures. More refined models incorporating mechanistic considerations, usually with reference to the multistage action of the carcinogenic process, have been proposed (for example, Siemiatycki & Thomas, 1981). Note that, under a multistage hypothesis, one would normally expect two carcinogens that act on different stages to act multiplicatively, whereas two carcinogens acting on the same stage might act additively.

The additive and multiplicative models outlined above have the advantage that they can also be formulated in terms of relative risks. Let $R_{ij} = p_{ij}/p_{00}$ $[(i, j) \neq (0, 0)]$ be the relative risk of group (i, j) compared with the control group, where neither exposure is present. The additive model then predicts that $R_{ij} = R_{i0} + R_{0j} - 1$, whereas the multiplicative model predicts that $R_{ij} = R_{i0}R_{0j}$.

Formulating these models in terms of relative risks enables utilization of the basic methods for the analysis of long-term animal experiments described in Chapter 5. These methods, which account for intercurrent mortality and consider the context of observation of tumours, allow one to describe the differences in tumour yield between two or several groups in terms of relative risks. We illustrate this by an example, based on our study of all combinations of $I + 1$ levels of A and $J + 1$ levels of B, in which we investigate the multiplicative model for the joint action of the two exposures.

If a_i is a fixed level of exposure A, then the groups receiving the combination $(a_i, b_0) \cdots (a_i, b_J)$ form a subexperiment which can be analysed by the methods described in Chapter 5. The particular analysis may depend on the context of observation of the tumours and on whether time to death is available, but in all cases one should calculate for each level of factor B an expected number of tumour-bearing animals E_{a_ij} which can be contrasted with the respective observed numbers O_{ij}.

$$R_{a_ij} = (O_{ij}/E_{a_ij})/(O_{i0}/E_{a_i0})$$

then represent the relative risks at the different dose levels of factor B with reference to the baseline $b_0 = 0$. Such an analysis can be carried out at all $r + 1$ levels of factor A, and when the effect of factor B is the same at all these levels, it is justified to summarize it by adding the O's and E's over the different levels of factor A.

$$O_{.j} = \sum_{i=0}^{I} O_{ij} \quad \text{and} \quad E_{.j} = \sum_{i=0}^{I} E_{a_ij}$$

then denote the so-derived summary values, and the relative risks at the different levels of factor B, averaged over factor A, would then be

$$R_{.j} = (O_{.j}/E_{.j})/(O_{.0}/E_{.0}).$$

This process of averaging the effect of one factor over the different levels of the other assumes that the relative risks are the same, irrespective of which level of factor A is considered. This means that the assumption of a multiplicative model (as formulated above) is made implicitly.

In exactly the same way as we have summarized the effect of factor B averaged over factor A, we can derive a summary description of the effect of factor A averaged over factor B. This yields the observed and expected numbers $O_{i.}$ and $E_{i.}$ and hence the summary relative risks

$$R_{i.} = (O_{i.}/E_{i.})/(O_{0.}/E_{0.}).$$

Finally, we can also calculate the relative risks of developing a tumour in any of the single combination groups compared to the untreated control group. This is done by conducting an analysis only with the particular group of interest (a_i, b_j) and the untreated control group (a_0, b_0). If we denote the observed and expected numbers in this analysis as O_{ij}, O_{00}, E_{ij} and $E_{00}^{(i,j)}$, the resulting relative risks are then

$$R_{ij} = (O_{ij}/E_{ij})/(O_{00}/E_{00}^{(i,j)}).$$

Under the multiplicative model, one would expect to observe that

$$R_{ij} = R_{i.}R_{.j}.$$

This can be inspected in an informal way, with the calculated relative risks. Either the model is reasonably fulfilled for all $r \cdot s$ combination groups, or the pattern of deviation will provide an indication of whether the multiplicative model applied fits the data or not. It should be made clear that this approach does not represent a full-scale fitting of a statistical model, as the random variation behind the relative risk estimates is not considered in this descriptive approach. Also, the method is likely to be useful only when the observed number of tumours in the untreated control group is not too small, otherwise the variation in relative risk will be very large. However, this descriptive approach can give an indication of possible underlying models (Métivier et al., 1984). It can be used to investigate whether a multiplicative model for relative risks of life-time development of tumours can explain the joint effect. Adjustment for intercurrent mortality and context of observation is achieved by utilizing the methods given in Section 5.7.

For the analysis of observable tumours or rapidly lethal tumours, survival methods can be applied, and the probability of tumour-free survival to the end of the experiment may be a relevant endpoint to consider. A multiplicative model for this effect measure would imply that the age-specific hazard in a combination group is the sum of the age-specific hazards of the respective single-exposure groups. This was shown by Wahrendorf *et al.* (1981), and Korn and Liu (1983) proposed likelihood ratio tests for this purpose.

To assume that the age-specific hazard rates are multiplicatively related in the absence of an interaction has been used in the framework of Weibull models (see Section 6.3). Assuming common values for w and κ, the parameters β_{ij} fitted for each group can be viewed as relative hazard rate parameters. A formal test of the multiplicative model for the hazards can be achieved by using maximum likelihood methods (Peto & Lee, 1973) to compare the models $\log \beta_{ij} = \mu + t_i + c_j$, which may be called the multiplicative main effects model, and $\log \beta_{ij} = \mu + x_{ij}$, which is a saturated model. This results in a chi-squared statistic on IJ degrees of freedom testing the overall fit of the multiplicative main effects model. Departures from the fit can be investigated further by comparing the observed number of animals with tumours in each group O_{ij} with that expected under the first model

$$\hat{E}_{ij} = \hat{\beta}_{ij} v_{ij},$$

where v is defined in Section 6.3.

An analogous analysis of time to tumour or death from tumour can be based on the proportional hazards model (Section 6.3). In this case one would assume that the hazard function satisfies the equation

$$\lambda_{ij}(t) = \exp(\Delta_{ij})\lambda_0(t)$$

for (i, j), where postulating $\Delta_{ij} = \mu + t_i + c_j$ would lead to a multiplicative model for the hazard functions, and where analogous comparisons between the multiplicative main effects model and the saturated model can be performed. More formal tests of the multiplicative model can be carried out in a straightforward manner when time to death can be ignored and the data can be expressed as simple counts of tumour-bearing animals in a $2 \times (I + 1) \times (J + 1)$ contingency table (Bishop *et al.*, 1974; Baker & Nelder, 1978).

Finally, it should be noted that studying quantal responses to mixtures of drugs has a long tradition in investigations of acute toxic effects. The different models considered in this framework have been reviewed by Hewlett and Plackett (1979, Chapter 7).

7.6 Litter effects

The statistical methods discussed in Chapter 5 are based on the assumption that the responses for different animals are statistically independent. As noted in Chapter 3, however, the assumption of independence may be violated with experimental designs involving the use of littermates, since pups within the same litter may tend to respond similarly (Gaylor *et al.*, 1985b). There are three main possibilities for distributing littermates among the experimental groups. First, animals from each complete litter may be assigned to the same group, as is necessary in two-generation experiments when the parental generation is exposed to different doses of a substance. Second, littermates may be distributed among different treatment groups in blocked designs with blocks defined in terms of equal-sized litters. Third, animals may be allocated to different treatment groups regardless of litter membership as in the completely randomized design.

In the first two cases, special methods of statistical analysis are required. In the first situation, it is necessary to take into account the within-litter correlation by using the variation between litters rather than variation between animals as the basis for between-group comparisons. Below, we discuss the statistical consequences of this in general terms and outline specific methods of analysis that may be used with such data. The third case, in which littermates are distributed across experimental groups, does not have as great an impact on the statistical inference and will be discussed at the end of this section.

Complete litters assigned to different groups

In the presence of positive intralitter correlation, standard methods of statistical analysis which ignore the litter structure will tend to underestimate the standard error of the difference between the overall response rates in two different treatment groups and hence overstate the statistical significance of any observed differences (Haseman & Hogan, 1975). Gladen (1979) showed that the use of standard chi-squared or likelihood ratio tests which ignore intralitter correlation can result in inflated false-positive rates. In this regard, Rao and Scott (1981) have shown that the correct asymptotic null distribution of the usual chi-squared statistic for comparing several treatment groups is in fact a weighted sum of independent chi-square random variables with weights related to the intralitter correlation within each group. The use of the Fisher–Irwin exact test for comparing two treatment groups, ignoring the litter structure, has also been shown to result in somewhat inflated false-positive rates in the presence of positive intralitter correlation. With negative intralitter correlation, however, the standard tests would be valid in the sense that the actual false-positive rate would tend to be less than the nominal rate.

In order to avoid these problems, statistical methods which take litter structure into account should be employed. Let n_{ij} denote the size of the jth litter in the ith treatment group $(j = 1, \ldots, m_i; i = 1, \ldots, I)$, and let x_{ij} denote the number of these animals developing tumours. Conditional on the n_{ij}, Cochran (1943) assumed that the sample proportions $\hat{p}_{ij} = x_{ij}/n_{ij}$ have mean

$$E(\hat{p}_{ij} \mid p_{ij}) = p_{ij}$$

and variance

$$V(\hat{p}_{ij} \mid p_{ij}) = p_{ij}(1 - p_{ij})/n_{ij},$$

where the p_{ij} are considered to be held constant. If the p_{ij} are actually independent random variables with mean

$$E(p_{ij}) = \mu_i$$

and variance

$$V(p_{ij}) = \sigma_i^2 > 0,$$

we have

$$E(\hat{p}_{ij}) = \mu_i$$

and

$$V(\hat{p}_{ij}) = \frac{\mu_i(1 - \mu_i)}{n_{ij}} + \sigma_i^2\left(1 - \frac{1}{n_{ij}}\right).$$

Note that the first term in the above formula represents the variation that would occur in the absence of any litter effects, whereas the second term reflects the between-litter variation associated with such effects. Cochran proposed a weighted analysis of variance of the observed proportions $\hat{p}_{ij}$ as a means of assessing treatment differences, with estimates of the litter-specific variances σ_{ij}^2 used to obtain the weights (see also Kleinman, 1973).

Gladen (1979) used a more general model with $V(\hat{p}_{ij}) = f(n_{ij})$ for some general function f. Although the natural estimator

$$\hat{p}_i = x_{i.}/n_{i.} = \sum_{j=1}^{m_i} x_{ij} \bigg/ \sum_{j=1}^{m_i} n_{ij}$$

of μ_i is unbiased, Gladen proposed the unbiased jackknife estimator

$$\hat{p}_{Ji} = m_i\hat{p}_i - \frac{(m_i - 1)}{m_i} \sum_{j=1}^{m_i} \hat{p}_{i(-j)},$$

where $\hat{p}_{i(-j)} = (x_{i.} - x_{ij})/(n_{i.} - n_{ij})$ denotes the estimator of μ_i omitting the jth litter in group i. The jackknife estimator of the variance of each $\hat{p}_{Ji}$ is given by

$$v_{Ji} = \frac{m_i - 1}{m_i} \sum_{j=1}^{m_i} \left[\hat{p}_{i(-j)} - \frac{1}{m_i} \sum_{j=1}^{m_i} \hat{p}_{i(-j)}\right]^2.$$

To compare two groups with m_1 and m_2 litters, respectively, the statistic

$$t_J = \frac{\hat{p}_{J1} - \hat{p}_{J2}}{(v_{J1} + v_{J2})^{\frac{1}{2}}}$$

will then approximate a t-distribution with $m_1 + m_2 - 2$ degrees of freedom under the null hypothesis $H_0: \mu_1 = \mu_2$, provided m_1 and m_2 are sufficiently large.

Frangos and Stone (1984) investigated, among other approaches, the estimator

$$\bar{p}_i = \sum_{j=1}^{m_i} \hat{p}_{ij}/m_i,$$

which is the average of the litter-specific proportions in group i. The variance of $\bar{p}_i$ can be estimated by

$$\bar{v}_i = \sum_{j=1}^{m_i} (\hat{p}_{ij} - \bar{p}_i)^2/\{m_i(m_i - 1)\}.$$

For comparison of two groups with m_1 and m_2 litters, this would lead to a standardized test statistic

$$z = (\bar{p}_1 - \bar{p}_2)/(\bar{v}_1 + \bar{v}_2)^{\frac{1}{2}},$$

which asymptotically follows a standard normal distribution. The small sample behaviour of z has not been investigated. For a single group, however, Frangos and Stone (1984) demonstrated that confidence intervals based on $\bar{p}_i$ and on a modification of an estimator proposed by Southward and Van Ryzin (1972) outperform the jackknife confidence intervals.

Another nonparametric technique has been considered by Haseman and Soares (1976). In particular, they showed that comparing the litter-specific proportions in two treatment groups using the nonparametric Wilcoxon test (Hollander & Wolfe, 1973, p. 68) is in many circumstances a sufficiently accurate and powerful statistical procedure.

A more parametric approach to modelling litter effects was used by Williams (1975), who assumed that p_{ij} follows a beta distribution with parameters $\alpha_i, \beta_i > 0$. In this case, $\mu_i = \alpha_i/(\alpha_i + \beta_i)$ and $\sigma_i^2 = \mu_i(1 - \mu_i)\rho_i/(1 + \rho_i)$, where $\rho_i = (\alpha_i + \beta_i)^{-1} > 0$ provides a measure of the degree of association between litter mates. Williams proposed the use of beta-binomial likelihood ratio methods to test the null hypothesis $H_0: \mu_1 = \mu_2$; $\rho_1 = \rho_2$ against a general alternative. Williams also considered $H_0: \mu_1 = \mu_2 = \mu$ with $\rho_1 = \rho_2 = \rho$ fixed. Based on a simulation study, Shirley and Hickling (1981) concluded that, for typically encountered litter sizes, the nonparametric Wilcoxon test was preferable to the beta-binomial likelihood ratio test. (See also Haseman & Kupper, 1979.)

A different approach has been suggested by Kupper and Haseman (1978). They assume that 'fetuses in the same litter tend to have an inherent relationship to one another.' Thus, the assumption of mutual independence of the outcomes within a litter, which usually leads to a binomial within-litter model, is altered by applying a correction factor which depends on the covariance between two Bernoulli trials within one litter. Kupper and Haseman (1978) demonstrate that this correlated binomial model is in a sense an extension of the beta-binomial model, as discussed by Williams (1975), in that it also allows, to some degree, negative correlations between responses within a litter. A likelihood ratio test is again proposed to test for differences between the experimental groups. For one example, Kupper and Haseman demonstrated a better fit of their correlated binomial model than of the beta-binomial model.

All of the procedures proposed above are based on asymptotic approximation and rely on large m_i for their validity. This is particularly important when the response probabilities μ_i are near zero or one. Because many rodent lesions occur with frequencies of 1% or lower, exact permutation tests of the null hypothesis $H_0: \mu_1 = \mu_2$ against the alternative $H_2: \mu_1 > \mu_2$ based on the observed difference $d = \hat{p}_1 - \hat{p}_2$ may be considered (Crump & Howe, 1980). These tests are based on the fact that, under the null hypothesis, each of the $s = \binom{m_1 + m_2}{m_1}$ possible assignments of m_1 litters to group one and m_2 litters to group two are equally likely. The significance level for the exact randomization test against the alternative $H_1: \mu_1 > \mu_2$ is then given by r/s, where r is the number of permutations leading to a value of d at least as large as the observed value. When s is large, Crump and Howe suggest that the significance level may be estimated on the basis of a random sample from the permutation distribution. In the special case $n_{ij} \equiv n$, the algorithm given by Soms (1977) may be used to obtain the randomization significance level. Although these procedures are exact, they are conservative for small values of μ in the sense that the false-positive rate can be notably less than the nominal level (Krewski *et al.*, 1984a).

Several approaches to modelling dose-response data which take into account intralitter correlation have also been proposed. Under the beta-binomial model considered by Williams (1975), the marginal distribution of x_{ij} (the number of animals developing tumours in the jth litter in the ith treatment group) is a beta-binomial, so that the likelihood for the parameters $\mu_i = \alpha_i/(\alpha_i + \beta_i)$ and $\rho_i = (\alpha_i + \beta_i)^{-1} > 0$ ($i = 1, \ldots, I$) is a product of I independent beta-binomial terms. Segreti and Munson (1981) then proposed that the effects in dose d_i administered to the ith group be modelled as

$$\mu_i = \lambda + (1 - \lambda)F(\alpha + \beta \log d_i),$$

where $0 < \lambda < 1$ and $\beta > 0$ as in (6.2) and (6.8). The beta-binomial likelihood may then be used to obtain estimates of the parameters α, β and λ in the presence of dose-specific litter effect parameters $\rho_1, \ldots, \rho_I$. (Segreti and Munson also consider a simpler but less realistic model in which $\rho_1 = \cdots = \rho_I = \rho$.) The former three estimates then provide a fitted dose-response curve

$$\hat{\mu} = \hat{\lambda} + (1 - \hat{\lambda})F(\hat{\alpha} + \hat{\beta} \log d).$$

Another approach to this problem has been studied by Ochi and Prentice (1984). In general terms, they consider a correlated probit regression model in which the binary responses within the same litter are defined as indicators of whether or not the corresponding components of a multivariate normal regression vector with common mean and variance exceed some threshold value. Although the likelihood calculations are somewhat more complex than in the Segreti–Munson model, the Ochi–Prentice model provides for multiple covariates as well as flexibility in modelling changes in intralitter correlation with dose.

Litters distributed across groups in blocked designs

The situation in which littermates are distributed across litters has been investigated by Mantel *et al.* (1977) and by Mantel and Ciminera (1979). For a discussion of this issue see also Mantel (1980). Basically, they suggest comparing the tumour incidence in different treatment groups by stratifying over the litters. This could be done with the methods discussed in Chapter 5. However, as litters are usually not very large, it may be that, towards the end of an experiment in certain litters, animals in only one experimental group are at risk, with no surviving littermates for comparison. Thus, the information from these animals may remain unused. To avoid this, Mantel *et al.* (1977) suggest several special devices for combining remaining animals into new strata. Mantel and Ciminera (1979) outline a different approach, in which so-called 'Savage-scores' are assigned to each animal, irrespective of litter and group, based on when or whether the animal developed a tumour. Using these scores, which are common in life-table analyses, litter-adjusted comparisons between control and treated groups were proposed.

Problems with this approach were pointed out by Michalek and Mihalko (1983), with a discussion by Mantel (1983). It was demonstrated that the attempt to utilize remaining information can confound litter and treatment effects. It should also be noted that use of the hypergeometric variance, as in Chapter 5 [formula (5.2)], is preferable for stratified comparisons to the permutational variance used by Michalek and Mihalko (1983) because the permutational variance is invalid when treatment influences mortality. Therefore, a simple litter-stratified analysis, as outlined by Mantel *et al.* (1977) and Michalek and Mihalko (1984), but without recovery of interlitter information, appears to be the most advisable approach. In any case, complete randomization of experimental animals into all the experimental groups, irrespective of their litter membership, represents the preferable experimental design.

7.7 Association among tumour types

It is obvious from the preceding chapters and sections that several tumour types are investigated in a long-term animal experiment. Statistical analysis is usually performed for each of these tumour types individually. This may lead to problems of multiple comparisons in making statistical inferences from the study, as discussed in Section 7.2. The association of a given tumour type with another represents in this context a nuisance factor which is manifested in the intercurrent mortality. Methods accounting for this are discussed at length in Chapter 5.

However, the association among tumour types also represents an interesting aspect of studies on the mechanism of action of the exposure in the entire biological system (for example, animal) investigated. For the moment we shall neglect the role of treatment and consider only animals from one group. Looking at two tumour types, A and B, say, one could define the association between these tumour types by the odds ratio in the resulting 2×2 table if one categorizes the animals according to the occurrence of the two tumours of interest:

		Tumour A	
		absent	present
Tumour B	absent	a	b
	present	c	d

The association would be defined as $\hat{\psi} = (ad)/(bc)$, and standard statistical methods for odds ratios (see Chapter 5) could be employed. However, before doing so, careful attention must be given to the way the tumours have been found in the animals. The intercurrent mortality, probably influenced by the presence of one or both of the two tumours, plays a crucial role. Consider the simple model of Breslow *et al.* (1974).

Assume X_A and X_B to be the times of clinical onset of tumours A and B, defined operationally as the earliest time the tumours would be detected by necropsy. Let Y_A and Y_B be the times from onset until death from tumours A or B, and Z the time of death due to an unrelated cause, including serial sacrifice. An animal would be classified in one of the cells of the above 2×2 table if

   (a) $Z < \min(X_A, X_B)$: neither tumour present at necropsy;
   (b) $X_A < X_B$ and $\min(Z, X_A + Y_A) < X_B$: only tumour A present;
   (c) $X_B < X_A$ and $\min(Z, X_B + Y_B) < X_A$: only tumour B present;
   (d) $\max(X_A, X_B) < \min(X_A + Y_A, X_B + Y_B, Z)$: both tumours present.

Assume that there is no association between tumours A and B, that is, X_A and X_B are independent random variables, but one tumour, say A, is rapidly lethal, that is, Y_A is very small. Then animals with both tumours present are very unlikely to be found among those dying. Based on such data the estimated association would appear to be unjustifiably negative, that is, $\hat{\psi} < 1$. However, if one constructed a 2×2 table, as above, on the basis only of animals which died from causes not related to the tumour(s), such as by serial sacrifice, the rapid lethality of one tumour would lead to the general finding of a low proportion of animals with this tumour, either alone or in combination with the other tumour. Therefore, the resulting estimate of the association should not have a systematic error.

The method for the analysis of carcinogenicity data proposed by Turnbull and Mitchell (1978) and Mitchell and Turnbull (1979), as already discussed in Section 6.3, includes prevalence models. Tumour prevalence can be observed directly only by serial sacrifice of animals. Therefore, such designs are needed to obtain an unbiased assessment of the association among tumour types. The log-linear prevalence model (see Section 6.3) can depend on various factors. These factors may include aspects of the experimental design, such as treatment group or time period of scheduled sacrifice, but also relate to the presence or absence of certain tumours. The resulting interaction terms between different tumour types can be used to evaluate possible associations. In addition, the interaction terms of each individual tumour with the treatment group will indicate whether the occurrence of this tumour depends on the treatment of the animals. Also, the interaction terms with the time periods will give indications of the temporal pattern of the tumour prevalence. This log-linear model for prevalence is combined with a logistic model for lethality which potentially depends on the same set of factors and interactions, but for which a different subset may prove to be significant. Interpretation of the results from both models has to be made jointly and the results have to be checked carefully for biological consistency.

In the search for jointly best-fitting prevalance and lethality models, the same approaches used for the fitting of multiplicative models for discrete data (Bishop *et al.*, 1974) can be applied. The operational criteria by which interaction terms are successively inserted or deleted may vary from application to application. Usually, likelihood ratio statistics are employed to judge whether a significant change in the goodness of fit is observed when altering the model in one direction or another. An example using this approach, but also including consideration of the simple 2×2 tables above, has been given by Wahrendorf (1983). Data from a long-term carcinogenicity study with DDT using CF-1 mice were used. In this study, which was originally reported by Tomatis *et al.* (1974), mice were fed 250 ppm of DDT for 15 or 30 weeks, after which exposure ceased. An untreated control group was also used. Scheduled sacrifices were conducted at 15, 30, 65, 95 and 120 weeks of exposure, though not equally frequently in all groups. Consequently, only two time intervals were available to define the corresponding factor in the prevalence and lethality model.

Three tumour types were investigated: lymphomas, liver and lung tumours. In the prevalance model, liver tumours showed a significant interaction with the factor treatment group, demonstrating the well-known hepatocarcinogenic effect of DDT. The prevalence of all three tumours showed a significant interaction with time, indicating increased occurrence of all tumours in the later time interval. However, there also remained significant negative interaction terms between lymphomas and liver tumours and between lymphomas and lung tumours. These were inspected further by calculating coefficients of association in a simple 2×2 table contrasting two tumours. Such tables were derived by using only those animals which were sacrificed in each group and each time interval. This showed a consistent pattern of negative association between lymphomas and liver tumours. Counteracting this negative association, by including among those animals with both tumours also those who died naturally, did not alter this conclusion. As the prevalence of lymphomas was not related to treatment group, it was concluded that the hepatocarcinogenic activity of DDT may have an influence on the development of lymphomas in CF-1 mice.

7.8 Historical control tumour rates

In the evaluation of a chemical carcinogenesis experiment, knowledge of the spontaneous tumour rates obtained from control groups of previous experiments can often provide insight into the possible carcinogenicity of a test compound (Gart *et al.*, 1979; Tarone *et al.*, 1981; Haseman, 1983a). The most appropriate and important comparison of an exposed group is with the control group randomized from the same source. However, historical control tumour rates can be helpful in evaluating experiments for which the statistical analysis based on matched control tumour rates indicates equivocal evidence of carcinogenicity. One situation in which historical rates are likely to be particularly helpful is in the evaluation of small nonsignificant tumour increases at tissue sites with very low spontaneous tumour rates. When historical control rates are used to evaluate an equivocal experiment, both the magnitude and variability of these rates must be considered (Tarone *et al.*, 1981; Haseman, 1983a).

Although informal, ad-hoc comparisons with historical control data can often provide some insight into the carcinogenic potential of a test chemical (Fears *et al.*, 1977; Tarone *et al.*, 1981), methods have recently been developed which permit the incorporation of historical control information in a formal framework. Tarone (1982) modelled historical control rates using a beta-binomial model and derived a test for dose-related trends which is a modification of the Cochran–Armitage test. The modification depends both on the magnitude and variability of the historical rates. Hoel (1983) proposed an exact test based on the beta-binomial model. When the parameters of the beta-binomial distribution are known, Hoel's exact test is valid, and

Tarone's test is asymptotically valid (Krewski *et al.*, 1985; Hoel & Yanagawa, 1986). Problems arise, however, when the parameters must be estimated from the available historical data (Tamura & Young, 1986). Bias in the estimates of the beta-binomial parameters causes the methods to give too much weight to the historical control data. Thus, methods based on the beta-binomial model should be used with caution until unbiased estimators of the beta-binomial parameters are developed.

Dempster *et al.* (1983) assume that the logits of the historical control rates are normally distributed, and evaluate the evidence of a dose-response relationship using a Bayesian analysis, again incorporating information about the magnitude and variability of the historical rates. Dempster *et al.* also discuss diagnostic methods to assess the sensitivity of their analysis to different prior distributions and to assess the goodness of fit of the various models (including the beta-binomial model). The small-sample performance of the method of Dempster *et al.* has not been investigated, but, because of the tractability of estimation procedures for normal models, it is unlikely that their method will share the problems associated with those based on the beta-binomial model.

In making a formal analysis based on historical data, care must be taken to ensure that the historical control rates used in the analysis come from experiments which are similar to the current experiment in factors known to affect the magnitude of spontaneous tumour rates. Such factors may include the length of time on study, housing conditions, type of food, and possibly the year of birth of the test animals (Gart *et al.*, 1979; Tarone, 1982; Haseman, 1983a). Certainly, some initial screening is necessary to determine which historical rates may be used in the analysis of a particular experiment. In cases where the historical control data are informative with respect to the concurrent control response rate, their use may greatly strengthen the inferences made concerning the hypothesis of carcinogenicity. In contrast to their value in hypothesis testing, however, historical control data seem to provide little additional information when modelling dose-response relationships (Smythe *et al.*, 1986).

8. ANALYSIS OF AUXILIARY DATA

CHAPTER 8

ANALYSIS OF AUXILIARY DATA

8.1 Introduction

In most long-term animal carcinogenicity studies, data are acquired on many variables other than neoplastic and non-neoplastic lesions. As noted in Chapter 5, for example, individual survival times may be of use both in establishing differences in mortality patterns among the various treatment groups and in adjusting for such differences in comparisons of tumour occurrence patterns between groups. Other key variables routinely monitored in long-term studies include body weight and feed consumption, clinical signs of toxicity, haematological parameters and organ weights taken at time of necropsy.

Statistical analysis of such auxiliary data may be broadly categorized into one of three general types. A variety of established procedures for the analysis of censored failure-time data can be used for survival data (Section 8.2). Continuous variables monitored at a particular point in time, such as terminal organ weights or body weight at 12 months on test, may be dealt with using analysis of variance procedures (Section 8.3). Variables observed at successive points in time, such as weekly body weight, are subject to repeated measures or growth-curve analyses (Section 8.4). The remainder of this chapter provides an overview of statistical techniques available within each of these three categories.

The analysis of concomitant information is not the main goal of the statistical analysis of a long-term experiment but assists in interpreting the findings with respect to carcinogenicity. The particular methods addressing carcinogenicity form the main part of this book and have been discussed in the preceding chapters. In this chapter, we shall give only a brief introduction to the variety of techniques available to analyse auxiliary data. It should be made clear that the methods mentioned in this chapter are generally not suitable for the analysis of carcinogenicity.

8.2 Analysis of survival data

It has become apparent throughout this monograph that mortality plays an important role in evaluating carcinogenicity in long-term animal experiments. Before any evaluation of the carcinogenic response is undertaken, a thorough examination of the underlying survival pattern should be performed. This will identify differential mortality patterns that can lead to bias in the assessment of the carcinogenic response. Thus, survival analysis can assist in the choice of appropriate methods to adjust for differences in intercurrent mortality. The particular methods for the analysis of survival data have already been introduced in Chapter 5. In this section, we give a brief summary of these methods, with cross references to the appropriate sections.

The most common approach to the analysis of survival data in a long-term animal experiment is to estimate the survival curves in each experimental group. These are then displayed graphically, and statistical tests are performed to find whether there are significant differences in survival among the experimental groups or whether there is a significant trend in survival with increasing dose. Methods appropriate to these issues are outlined and illustrated in Section 5.3. The impact of different survival patterns on the assessment of the carcinogenic response is discussed at length in Chapter 2, specifically in Table 2.2. Methods adjusting for differences in survival in the analysis of carcinogenicity are discussed in detail in Sections 5.5, 5.6 and 5.7.

If survival as such appears to be an endpoint which merits more detailed analysis, for example by regression analysis to study the effect of other covariates apart from dose, the proportional hazards model introduced in Section 6.3 is the method of choice. This flexible regression model is a natural extension of the log-rank test for comparing survival in several groups, given in Section 5.3. Furthermore, the proportional hazards model allows the investigation of time-dependent covariates. For example, the influence on an animal's survival of its body weight (if monitored continuously during the experiment) could be analysed using the proportional hazards model. Proportional hazards methods are discussed by Kalbfleisch and Prentice (1980, Chapter 5), Miller (1981b, Chapter 6) and Cox and Oakes (1984, Chapter 8). These authors also provide a thorough treatment of all statistical issues of survival analysis, whereas Lee (1980) provides a more elementary text.

8.3 Analysis of variance

Consider a simple experiment in which there are I treated groups exposed to doses $d_1 < \cdots < d_I$ and an unexposed control, with dose $d_0 = 0$. The manner in which animals are assigned to various treatment groups and the manner in which the experiment is conducted will determine the appropriate analysis for the experiment at hand. As discussed in Chapter 3, the animals should be randomly assigned to each dose in accordance with the experimental design.

The simplest possible randomization scheme is to assign the available animals to the various treatment groups completely at random. With only one animal housed in each cage, this leads to the completely randomized design, discussed in Chapter 3. The familiar randomized block design means that the animals are grouped into a number of homogeneous blocks prior to randomization (for example, on the basis of initial body weight or litter status), with animals from each block randomly assigned to each treatment. In this case, the blocking factor (initial body weight or litter status) must be taken into account in the analysis of variance. Even with complete randomization, the conduct of the experiment is important in determining the method of statistical analysis. With two animals housed in each cage, for example, any cage effects are 'nested' within treatment effects and should be considered in the analysis-of-variance model employed.

There are two major categories of statistical methods available for the analysis of experimental data: parametric methods, which are based on specific assumptions (usually normally distributed data with equal variances in each group), and nonparametric methods, which are not based on such assumptions and often replace the actual observations by their ranks.

In describing these methods, we follow closely two main textbooks – one on parametric methods (Brownlee, 1965) and the other on nonparametric methods (Hollander & Wolfe, 1973). Many other textbooks also provide a good coverage of these methods and could be used when studying technical aspects in detail. The introductory nature of this section requires restriction to essential principles, and it should be borne in mind that the analysis of concomitant information should help in interpreting the findings of a long-term carcinogenicity study but does not, in general, play a central role.

Parametric methods

Let y_{ij} denote the response of animal j ($j = 1, \ldots, n_i$) at dose i ($i = 0, 1, \ldots, I$). As noted earlier, y_{ij} might represent body weight, feed consumption or any other continuous variable observed at a specified point in time. Let

$$\bar{y}_i = \sum_{j=1}^{n_i} y_{ij}/n_i$$

denote the mean of the n_i observations at dose i and

$$\bar{y} = \sum_{i=0}^{I} \sum_{j=1}^{n_i} y_{ij}/n$$

denote the mean of the

$$n = \sum_{i=0}^{I} n_i$$

animals in the experiment. The standard analysis-of-variance model for the completely randomized design is formulated as

$$y_{ij} = \mu + \tau_i + \varepsilon_{ij}, \tag{8.1}$$

where μ is a constant, τ_i denote the effects of treatment $i = 0, \ldots, I$, with $\sum_{i=0}^{I} \tau_i = 0$, and ε_{ij} are random error terms assumed to be independent, identically-distributed, normal random variables with a mean 0 and variance σ_i^2.

The assumption that the ε_{ij}'s have a normal distribution can be checked using a normal probability plot of residuals (Daniel & Wood, 1971) or using a goodness-of-fit test (Sokal & Rohlf, 1981, p. 696; Miller, 1986, p. 82). If the design is not badly unbalanced (i.e., if the n_i do not vary greatly), then moderate departures from normality have very little effect on the nominal significance levels of the analysis of variance methods presented in this section (Miller, 1986, pp. 80–82). Likewise, inequality of error variances has little effect on the analysis of variance tests unless the design is badly unbalanced (Miller, 1986, pp. 89–92). A preliminary test of homogeneity of error variances is not, in general, recommended. Rather, if visual inspection reveals obvious heterogeneity of error variances, then steps should be taken to try to reduce that heterogeneity before applying analysis of variance methods (Miller, 1986, pp. 92–94).

Failure of the assumptions regarding normality and homogeneity of error variances could be due to the presence of anomalous values or outliers in the data. If these can be identified from the residual plots, they can be either corrected or eliminated prior to analysis of the data. In some cases, heterogeneity of variance can be avoided by using a suitable transformation of the data. If the variance σ_i^2 is proportional to the group mean $\bar{y}_i$, for example, the transformation $y' = \sqrt{y}$ will result in homogeneous error variances (Brownlee, 1965, p. 145). Generally, the Box–Cox power transformation (Box & Cox, 1964) can be employed in an attempt to achieve simultaneously both normality and homogeneity of variance.

In the usual one-way analysis of variance for the completely randomized design, the variability among the observed treatment group means is compared to the within-group variability using a standard F-test (Brownlee, 1965, p. 312). As indicated in Table 8.1, this involves calculation of a sum of squares, SS_D, between the treatment group means and a pooled within-treatment sum of squares for error, SS_E. After dividing by the degrees of freedom to form the corresponding mean squares MS_D and MS_E, the ratio $F = MS_D/MS_E$ follows a central F-distribution under the null hypothesis $H_0: \tau_i = 0$ $(i = 0, 1, \ldots, I)$.

Table 8.1 Analysis of variance for the completely randomized design

Source of variation	Degrees of freedom	Sum of squares	Mean square	Expected mean square	F statistic
Dose	I	$SS_D = \sum_{i=0}^{I} n_i(\bar{y}_i - \bar{y})^2$	$MS_D = SS_D/I$	$\sigma^2 + \sum_{i=0}^{I} n_i(\tau_i - \bar{\tau})^2$	MS_D/MS_E
Error	$n - I - 1$	$SS_E = \sum_{i=0}^{I} \sum_{j=1}^{n_i} (y_{ij} - \bar{y}_i)^2$	$MS_E = SS_E/(n - I - 1)$	σ^2	
Total	$n - 1$				

If the between-group variation is significantly higher than the within-group variation, then there is evidence of significant differences between the treatment effects τ_i. However, no indication of what these differences are is provided. For this reason, tests for trend or multiple comparison procedures, which are described below, can be informative.

For the following, we denote by

$$s_i^2 = \sum_{j=1}^{n_i} (y_{ij} - \bar{y}_i)^2/(n_i - 1)$$

the error mean square for treatment group $i = 0, 1, \ldots, I$ with $E(s_i^2) = \sigma_i^2$. The pooled error mean square is then given by

$$s^2 = \sum_{i=0}^{I} (n_i - 1)s_i^2 \bigg/ \sum_{i=0}^{I} (n_i - 1).$$

In what follows, we assume that $\sigma_i^2 = \sigma^2$ for $i = 0, 1, \ldots, I$, so that $E(s^2) = \sigma^2$. Note also that $s^2 = MS_E$ from Table 8.1.

Tests for trend

These have already been discussed in the framework of tumour data and survival (see Chapters 2 and 5) and can also be used with concomitant information. Monotonicity is represented by formulating the alternative $H_1: \tau_i \leq \tau_j$ $(0 \leq i < j \leq I)$, with at least one strict inequality. Monotone decreasing or two-sided alternatives may also be specified.

Armitage (1955) proposed a test for linear trend with equally spaced doses (represented here by the group index i) based on the regression model

$$y_{ij} = a + b \cdot i + \varepsilon_{ij}, \tag{8.2}$$

where a and b are parameters to be estimated. The null hypothesis $H_0: b = 0$ is rejected in favour of the alternative $H_1: b > 0$ if

$$\frac{\sum_{i=0}^{I} n_i(i - \bar{i})\bar{y}_i}{\sqrt{\sum_{i=0}^{I} n_i(i - \bar{i})^2}} \geq t_{\alpha, n-I-1} s,$$

where $\bar{i} = \sum_{i=0}^{I} i n_i / n$, and $t_{\alpha, n-I-1}$ denotes the $100(1 - \alpha)$ percentile of the t-distribution with $n - I - 1$ degrees of freedom. This test is identical to the test for significance of the slope in a linear regression except that s^2 is used to estimate σ^2 rather than the residual mean square error. This is done to eliminate any bias which would be included in the residual mean square for error if the true dose-response curve were not linear. Abelson and Tukey (1963) noted that for $n_i \equiv n_0$ $(i = 1, \ldots, I)$, that is, equal group sizes, Armitage's test is of the form

$$\frac{\sum_{i=0}^{I} c_i \bar{y}_i}{\sqrt{\sum_{i=0}^{I} c_i^2}} \geq t_{\alpha, n-I-1} \frac{s}{\sqrt{n}},$$

where $\sum_{i=0}^{I} c_i = 0$. Although it is impossible to choose the weights c_i to be uniformly most powerful against all possible monotone increasing functions, Abelson and Tukey suggested the weights

$$c_i = \left\{ (i-1)\left(1 - \frac{i-1}{I+1}\right) \right\}^{1/2} - \left\{ i\left(1 - \frac{i}{I+1}\right) \right\}^{1/2}.$$

Although Armitage's procedure will be more powerful if the dose-response curve is linear, there exist nonlinear alternatives for which the power of Armitage's test is smaller than the power of the Abelson–Tukey test. Extensions of this procedure to the case of unequal n_i are discussed by Barlow *et al.* (1972) and by Miller (1986, pp. 78–80).

Pairwise group comparisons

Although tests for trend are usually of greatest relevance and interest, it can sometimes be informative to carry out certain pairwise group comparisons (for example, comparing each of the I treatment groups to the control).

A test for the difference between any two groups (indexed, say, by h and i) can be performed by declaring the difference to be significant if

$$|\bar{y}_i - \bar{y}_h| \geq t_{\alpha/2, n-I-1} s \left(\frac{1}{n_i} + \frac{1}{n_h} \right)^{1/2},$$

(8.3)

where $t_{\alpha/2, n-I-1}$ denotes the $100(1 - \alpha/2)$ percentile of the t-distribution with $n - I - 1$ degrees of freedom and where α is the nominal significance level. This provides a valid test for the single comparison of group i to group h. However, if several such tests, say $M > 1$, are carried out (for example, comparing each treatment group in turn to the control, for which $M = I$), then the overall significance level (that is, the probability of finding at least one of the M tests significant under $H_0 : \tau_i = 0$ for all i) using the criterion in (8.3) will exceed the nominal level α. If, for example, two independent comparisons are performed with $\alpha = 0.05$, then the probability of declaring at least one of the two differences significant under H_0 is $1.0 - (1.0 - 0.05)^2 = 0.0975$, which is substantially higher than the nominal significance level, 0.05, of the two separate tests.

The goal of multiple-comparisons procedures is to allow several comparisions of interest to be made while maintaining the overall significance level at a fixed α. The methods presented require that the M comparisons to be made be chosen a priori. The simplest multiple comparisons method is the Bonferroni method (Miller, 1981a). The test criterion for the Bonferroni method is identical to that in (8.3) above, except that α is replaced by $\alpha' = \alpha/M$. A slight improvement on the Bonferroni method, particularly for large M, is provided by the Dunn–Sidák method (Dunn, 1974; Miller, 1981a), for which the test criterion is again identical to that in (8.3), but with α replaced by $\alpha'' = 1 - (1 - \alpha)^{1/M}$.

The comparisons that are likely a priori to be of general interest are of each treatment group in turn to the control. When $n_i = n_0$ for all i, the method of choice is the many-to-one t-test (Dunnett, 1955), which is performed by declaring a difference to be significant if

$$|\bar{y}_i - \bar{y}_0| \geq |d|_{\alpha, I, n-I-1} s \sqrt{2/n_0},$$

where $|d|_{\alpha, I, n-I-1}$ is tabulated by Dunnett (1955, Table 2). The many-to-one t-test for unequal sample sizes and tabulated critical values for the general case are described by Dunnett (1964) and Dutt et al. (1975, 1976).

Extensions of the one-way analysis of variance

Extension of the one-way analysis of variance may be required, since, as indicated in Chapter 3, many experiments do not follow the simple structure of a completely randomized design. For example, consider a two-generation study in which the parent or F_0 generation has been assigned to treatment groups in accordance with a completely randomized design, and the males and females in the same groups were mated on a one-to-one basis. Suppose now that a fixed number $m \geq 2$ of pups of each sex was selected from each litter to continue on test in the second or F_1 generation. Two animals from the same litter may be expected to have similar characteristics because of their common genealogy (see Chapter 3). Thus, this experiment has two levels of randomization, since the litters are first randomly assigned to treatments and then the pups are randomly selected from the litters. Here, the litter effect is considered to be nested within the main treatment effect.

As a second example, consider a single-generation experiment in which animals are assigned to treatments using a completely randomized design, but two or more animals are caged together. In this case, it is possible that animals housed together respond more similarly than animals housed in different cages. Thus, the cage effect is nested within the main treatment effect. These experiments can be analysed using the nested analysis of variance model

$$y_{ijk} = \mu + \tau_i + \lambda_{j(i)} + \varepsilon_{ijk}, \tag{8.4}$$

where y_{ijk} denotes the response for animal $k = 1, 2, \ldots, m_{ij}$, in litter or cage $j = 1$, $2, \ldots, n_i$, in treatment group $i = 0, 1, \ldots, I$. Here τ_i denotes the effect of treatment i, $\lambda_{j(i)}$ denotes the random effect of the jth level of the nesting factor within treatment i, and ε_{ijk} denotes a random error term. The ε_{ijk}'s are assumed to be independent normal random variables with mean 0 and variance σ_ε^2 and the $\lambda_{j(i)}$ are independent normal random variables with mean 0 and variance σ_λ^2.

If the m_{ij} are all equal, the analysis is straightforward. The litter or cage averages can be calculated and analysed using the procedures discussed previously for the randomized design. In addition, the significance of litter or cage effects can be assessed using standard analysis-of-variance procedures. If the number of animals varies from litter to litter, then procedures for analysis of the experiment are more complicated. Healy (1972) has proposed an analysis based on weighted averages of litter means, where the weights are estimated from the variance components. This procedure ignores the uncertainty in the estimation of the weights which may invalidate the technique for small sample sizes. Tietjen (1974) has examined a test for treatment effects based on a Satterthwaite approximation (Searle, 1971). However, the conventional F-test ignoring the imbalance appears to perform better than the Satterthwaite approximation.

Nested designs can be further generalized to the case where there are more than two levels of randomization. For example, consider a two-generation study in which males and females are assigned to treatments under a completely randomized design, with each male randomly paired with two females. This experiment may be viewed as having three levels of randomization, with sires randomly assigned to treatments, dams randomly allocated to sires for mating and pups randomly selected within the litters. The litters are thus nested within sires which are in turn nested within treatments.

Consider an experiment in which n animals are to be assigned to each of $I + 1$ experimental groups. Suppose that the $(I + 1)n$ animals are divided into n groups of size $I + 1$ so that all animals in the same group have similar weights, and that one animal from each weight group is randomly assigned to each treatment. The groups in this experiment are referred to as 'blocks', and the experiment is referred to as following a 'randomized complete block design'. The blocks do not have to be defined in terms of animal weight. For example, one could consider litters to be blocks and assign one animal from each litter to each treatment (see Section 7.5). The randomized complete block design can be analysed using the analysis of variance model

$$y_{ij} = \mu + \tau_i + \beta_j + \varepsilon_{ij}, \tag{8.5}$$

where y_{ij} denotes the response of the animal from block j given treatment i, μ is a constant, τ_i denotes the effect of treatment i, β_j denotes the effect of block j, and ε_{ij} is a random error term.

The above model is based on the assumption that each treatment has the same effect on each block. The effect of blocking is to remove from the error sum of squares a term which measures the variation in the observed response among blocks. It also reduces the experimental error of the estimated differences between treatments. If the inter-block variation is large, the randomized complete block design will provide more sensitive tests for treatment effects than the completely randomized design.

Nonparametric methods

The methods outlined above rely on the parametric assumption that the error terms in the respective analysis of variance models – (8.1), (8.2), (8.4) and (8.5) – are distributed according to a normal distribution with mean zero and some unknown variance. This assumption is frequently not met by the data one is analysing. Therefore, methods that make less stringent assumptions about the underlying distribution have been developed which can easily be employed, as they are simply based on ranks. When compared to the methods based on assumptions of normality, these methods have been shown to lose only slightly in efficiency when the assumptions are valid, but can be considerably more efficient when they do not hold.

We shall give a brief introduction to nonparametric methods which can be used to analyse continuous variables monitored at a particular point in time. This introduction follows the description of these methods in the textbook by Hollander and Wolfe (1973).

Let y_{ij} denote the response of animal j ($j = 1, \ldots, n_i$) to dose i ($i = 0, 1, \ldots, I$). We deal again with model (8.1), $y_{ij} = \mu + \tau_i + \varepsilon_{ij}$, but assume that the error terms ε_{ij} are mutually independent and follow some continuous random distribution. To test the null hypothesis, that all treatment effects τ_i ($i = 0, 1, \ldots, I$) are equal, against the alternative that they are not all equal, all n observations y_{ij} are ranked in ascending order, giving rank 1 to the lowest value and rank n to the largest. Let r_{ij} be the rank of observation y_{ij} in this joint ranking. The sum of ranks for observations in group i is

$$R_i = \sum_{i=1}^{n_i} r_{ij} \quad (i = 0, 1, \ldots, I),$$

the average rank in group i being denoted by $R_{i.} = R_i/n_i$. The average rank of all n observations is $R_{..} = (n + 1)/2$. In order to assess whether the ranks in the individual groups differ from the overall average, the statistic

$$H = \frac{12}{n(n + 1)} \sum_{i=0}^{I} n_i(R_{i.} - R_{..})^2$$

is computed. Under the null hypothesis of no difference between the treatment groups, H follows asymptotically a chi-square distribution with I degrees of freedom. This test is usually referred to as the Kruskal–Wallis test. For small sample sizes and limited numbers of groups ($I = 2$, $n_i \leq 5$), tables of exact critical values for H have been published (Hollander & Wolfe, 1973).

In the case of ties among the observations, the mean of the respective ranks, or midranks, may be assigned to all the tied observations. Consider g sets of tied observations and let t_j $(j = 1, \ldots, g)$ be their respective size. Then H should be corrected by dividing its value by

$$1 - \left\{ \sum_{j=1}^{g} (t_j^3 - t_j) \Big/ (n^3 - n) \right\}.$$

In the case of two groups only $(I = 1)$, the Kruskal–Wallis test is identical to the Mann–Whitney or Wilcoxon test.

Test for monotone trend

A test for monotone trend will be indicated in most experiments involving a series of increasing dose levels. The analysis of concomitant information must assess whether the observations of interest follow a corresponding trend. In addition, when multiplicity of tumours (Section 7.3) or graded responses (Section 7.4) are considered, nonparametric approaches to the analysis of these endpoints are indicated.

From the overall ranking used for the Kruskal–Wallis test, one test statistic for the presence of a positive trend with increasing dose can be derived as follows. The rank sum R_0 of the first group can be viewed as a two-sample Wilcoxon statistic, comparing the responses in group 0 with the pooled responses in groups 1 to I. Similarly, $R_0 + R_1$, the rank sum of groups 0 and 1 combined, can serve as a test statistic to compare these to groups jointly against the combined group 2 to I.

With $I + 1$ groups, I such two-sample comparisons can be considered. The sum of all their test statistics will be considered to test for the presence of a positive trend

$$L = R_0 + (R_0 + R_1) + \cdots + (R_0 + R_1 + \cdots + R_{I-1}) = \sum_{i=0}^{I-1} (I - i)R_i. \qquad (8.6)$$

Under the null hypothesis of no difference between the $I + 1$ groups the expectation of L is

$$E(L) = \frac{n+1}{2} \sum_{i=0}^{I-1} s_i,$$

where

$$s_i = \sum_{j=0}^{i} n_j$$

is the cumulative sample size up to, and including, group i. The variance of L is

$$\mathrm{var}(L) = \frac{n+1}{12} \left\{ \sum_{i=0}^{I-1} s_i(n - s_i) + 2 \sum_{i=0}^{I-2} \sum_{j=i+1}^{I-1} s_i(n - s_j) \right\}. \qquad (8.7)$$

Small values of L, that is, values below the expectation, are indicative of a positive trend. This leads to the following standardized test statistic

$$T_L = [E(L) - L] / [\mathrm{var}(L)]^{1/2}. \qquad (8.8)$$

Asymptotically, T_L follows a standard normal distribution. If the value of T_L exceeds z_α, the upper $(1 - \alpha)$ percentile of the standard normal distribution, a positive trend can be concluded with a significance level of α.

In the case of ties, midranks, r_{ij}^* say, are assigned and the test statistic T can be corrected by replacing, in formula (8.7), the term $(n + 1)/12$ by

$$\frac{1}{n(n-1)} \sum_{i=0}^{I} \sum_{j=1}^{n_i} \left(r_{ij}^* - \frac{n+1}{2} \right)^2.$$

This test, which is similar to the test proposed by Page (1963) for complete block designs and also described by Hollander and Wolfe (1973), has been proposed by Wahrendorf *et al.* (1985) for complete randomized designs in the framework of mutagenicity data. However, it is also perfectly applicable to the analysis of concomitant information or special responses in long-term animal experiments. Here, the consistency and strength of the trend can also be estimated by some nonparametric measure of the stochastic ordering between two populations.

As can be seen in (8.6), this nonparametric trend test weights the rank sums of all treatment groups by an integer score. For the many experiments conducted on a multiplicative dose scale, these scores correspond to the logarithms of the dose levels. Marascuilo and McSweeney (1967) have proposed a second test for trend where the actual dose levels are used as scores, and the construction of such a general rank test has also been noted by Cuzick (1985). These tests can easily be performed in a stratified situation by summing the differences $E(L) - L$ and the variances calculated according to (8.7) over the strata and then forming a standardized test statistic according to (8.8).

The above tests for trend are based on an overall ranking of the observations in all $I + 1$ groups. Another nonparametric test of trend, which is based on all $I(I + 1)/2$ pairwise comparisons of two groups, is the Jonckheere test (Jonckheere, 1954), also described by Hollander and Wolfe (1973). Two groups, u and v say $(u, v = 0, 1, \ldots, I; u \neq v)$, can also be compared by a Wilcoxon test by counting the number of pairs (α, β) for which $y_{u\alpha} < y_{v\beta}$. If $\phi(a, b) = 1$ if $a < b$, and 0 otherwise, this is

$$U_{uv} = \sum_{\alpha=1}^{n_u} \sum_{\beta=1}^{n_v} \phi(y_{u\alpha}, y_{v\beta}),$$

frequently referred to as Mann–Whitney counts. Summing these U_{uv} from all $I(I + 1)/2$ pairwise comparisons gives the Jonckheere statistic

$$J = \sum_{u=0}^{I-1} \sum_{v=u+1}^{I} U_{uv}.$$

Under the null hypothesis of no difference between the $I + 1$ experimental groups, this follows asymptotically a normal distribution with expectation

$$E(J) = \left\{ n^2 - \sum_{i=0}^{I} n_i^2 \right\} \Big/ 4$$

and variance

$$\text{var}(J) = \left\{ n^2(2n+3) - \sum_{i=0}^{I} n_i^2(2n_i+3) \right\} \Big/ 72.$$

In this case, large values of J are indicative of a positive trend, leading to the standardized test statistic

$$T_J = [J - E(J)]/[\text{var}(J)]^{1/2}$$

which, under the null hypothesis, follows a standard normal distribution.

Multiple comparisons

For the purpose of multiple comparisons, say M pairwise comparisons among groups (for example, comparison of each exposed group in turn to the control group), the average ranks of each group $R_{i.}$ $(i = 0, 1, \ldots, I)$ are used. These correspond to the group means used earlier in the parametric approach, and the underlying arguments regarding the logic of adjusting tests (because of the multiplicity of comparisons) are exactly the same as described there. We shall outline the approximation procedure given by Dunn (1964) and described by Hollander and Wolfe (1973).

Maintaining an overall significance level of α, one can decide that the response in group i is different from the response in group h, that is, $\tau_i \neq \tau_h$, if

$$|R_{i.} - R_{h.}| \geq z_{\alpha'} \left\{ \left(\frac{n(n+1)}{12} \right) \left(\frac{1}{n_i} + \frac{1}{n_h} \right) \right\}^{1/2},$$

where $z_{\alpha'}$ is the $100(1 - \alpha')$ percentile of the standard normal distribution and $\alpha' = \alpha/2M$. It has to be noted that the above comparison between a group i and a group h is based on the overall ranking of all observations, and, thus, it depends on the observations in the other groups. For detailed discussions, see Hollander and Wolfe (1973) and Miller (1981a).

8.4 Repeated measures and growth curves

A long-term study often involves repeated measurements of the same parameter in the same subject over a period of time. For example, blood samples may be taken from a subsample of animals at specified points in time and subjected to detailed haematological evaluation. Body weights are generally recorded for all animals in a study on a regular basis in order to establish growth profiles. Similar records of food consumption are also maintained.

Since early work on the analysis of such data by Box (1950), statistical methods for repeated measures of growth-curve data have undergone extensive development (Geisser, 1980; Woolson & Leeper, 1980). The essential difference between these procedures and those discussed in Section 8.2 is that repeated measures are taken on the same individual, and that one needs to allow for the possibility of correlation among these observations. In the first part of this section, we consider the use of multivariate linear models for this purpose. Nonparametric or related approaches to this same problem are then considered in the second part. Finally, some biologically-based growth models are described briefly.

Multivariate linear models

Let Y_{it} denote the measured value of a particular variable for individual $i = 1, \ldots, N$ at time $t = 1, \ldots, T$. These data may be conveniently summarized in matrix form as

$$\mathbf{Y} = \begin{bmatrix} Y_{11} & \cdots & Y_{1T} \\ \vdots & & \vdots \\ Y_{N1} & \cdots & Y_{NT} \end{bmatrix},$$

where each row corresponds to the set of results for one individual obtained during the course of the study period, and each column represents the results for all individuals at a particular point in time. A multivariate linear model for the data $\mathbf{Y}$ is then

$$\mathbf{Y} = \mathbf{X}\boldsymbol{\beta} + \mathbf{E}, \tag{8.9}$$

where $\mathbf{X}$ is an $N \times P$ design matrix consisting of zeros and ones indicating the treatment assigned to each individual and $\boldsymbol{\beta}$ is a $P \times T$ matrix of unknown parameters reflecting both treatment and time effects, with each column representing linear model regression coefficients for that time period. The rows $\mathbf{E}_i$ of the error matrix $\mathbf{E}$ are assumed to be independent, multivariate normal random variables with mean $\mathbf{0}$ and covariance matrix $\mathbf{S}_i$. Under this model, the rows $\mathbf{Y}_i$ are independent, multivariate normal random variables with mean $\mathbf{M}_i = (\mathbf{X}\boldsymbol{\beta})_i$ and covariance matrix $\mathbf{S}_i$. As for the univariate analysis of variance procedures discussed in the previous Section 8.3, a multivariate analysis of variance based on the model in (8.9) may be carried out under the homoscedasticity assumption that $\mathbf{S}_i = \mathbf{S}$ for all i. Details of this analysis are given by Morrison (1976) and in other texts on multivariate methods. The case of heteroscedasticity has been discussed by Chakravorti (1974). Procedures for handling missing values have been proposed by Kleinbaum (1973) and Leeper and Woolson (1982).

This multivariate approach to data on repeated measurements requires that sufficient data be available in order to estimate the many unknowns involved in the mean vector $\mathbf{X}\boldsymbol{\beta}$ and the dispersion matrix $\mathbf{S}$ of the data $\mathbf{Y}$. To reduce the dimensionality of $\boldsymbol{\beta}$, Potthoff and Roy (1964) suggested the use of polynomials in time t to provide for longitudinal effects (see also Khatri, 1966). The use of a low-order polynomial of degree $Q < T$, for example, would reduce drastically the number of parameters to be estimated when T is large. (This approach has been employed in the analysis of growth-curve data from a long-term bioassay of *ortho*-toluenesulfonamide conducted by Arnold *et al.*, 1980.) Similarly, further assumptions could be made concerning the form of $\mathbf{S}$ (Grizzle & Allen, 1969), although oversimplification may result in an increase in false-positive rates (Boik, 1981; Elashoff, 1981; Schwertman *et al.*, 1981). A parametric approach in which the regression coefficients of the growth curves are considered as random variables and provide the basis for statistical inference has been proposed by Schach (1982).

Nonparametric methods and related approaches

As with the nonparametric methods discussed in Section 8.3 for the univariate linear model, there exist multivariate nonparametric methods which are again based on ranks and can be applied to the situation of repeated measurements (Bhapkar & Patterson, 1977). Koch *et al.* (1980) provide a comprehensive overview of the methods available for different situations. A specific application suitable to situations of partially incomplete observations is given by Koziol *et al.* (1981). Both these papers provide many further references.

Another approach to the analysis of growth curves would be to fit a certain parametric model to the shape of each individual growth curve, and to extract certain parameters or functionals from the fitted curves which are particularly relevant to the biological aspects of the assay. For example, the slope of the growth curve or an estimate of its second derivative (acceleration of growth), the area under the curve, the location or value of a maximum or minimum, or a categorization of the curve's profile may represent such measures derived for each animal from its growth curve. These measures for the different treatment groups can then be compared using the nonparametric techniques for one-way analysis of variance as given in Section 8.3. Thus, this approach, initially suggested by Wishart (1938) and considered by Prestele *et al.* (1979) and Haux (1985), reduces the multivariate data of repeated measurements to univariate comparison. This approach is very promising for the analysis of auxiliary data in carcinogenicity studies, since it is based on easily interpreted parameters, it can allow for different numbers of observations per animal and it does not rely on strong parametric assumptions.

Robust estimation procedures (Huber, 1981) may also be considered as a means of avoiding the parametric assumptions required in the multivariate linear model. Pendergast and Broffitt (1985), for example, considered the use of *M*-estimation for growth curve data. This approach is based on an arbitrary loss function chosen to be less sensitive to outlying values than the quadratic loss function on which the multivariate analysis of variance is based. It appears to provide a robust alternative to the latter analysis. Like the nonparametric methods based on ranks, however, this robustness is achieved only with considerably more computational effort.

Biological growth-curve models

The multivariate linear models discussed previously are purely statistical in nature and, while often providing an adequate description of growth-curve data, do not have an underlying biological basis. Sandland and McGilchrist (1979) consider models which allow for an initial period of rapid growth followed by a period of slower growth and then a levelling off or even a decline in body weight. The logistic model, for example, is based on the differential equation

$$\frac{dW(t)}{dt} = aW(t)[b - W(t)],$$

where $W(t)$ denotes the expected body weight at time t. These models are specified in terms of biologically meaningful parameters and predict the anticipated shape, although this does not necessarily imply that the model represents the correct underlying growth mechanism (Kowalski & Guire, 1974).

Other stochastic models may be based on autoregressive processes in which successive errors may be correlated (Glasbey, 1979), or on stochastic differential equations (Sandland & McGilchrist, 1979).

Models which relate body weight to food consumption have also been proposed (Daniel, 1983). This last approach has been used to distinguish between weight changes attributable to changes in food consumption and those resulting from alterations in metabolism.

9. REFERENCES

CHAPTER 9

REFERENCES

Abbott, W.S. (1925) A method of computing the effectiveness of an insecticide. *J. Econ. Entomol.*, **18**, 265–267

Abdelbasit, K.M. & Plackett, R.L. (1982) Experimental design for joint action. *Biometrics*, **38**, 171–179

Abelson, R.P. & Tukey, J.W. (1963) Efficient utilization of non-numerical information in quantitative analysis: General theory and the case of simple order. *Ann. Math. Stat.*, **34**, 1347–1369

Aitkin, M. & Clayton, D. (1980) The fitting of exponential, Weibull and extreme value distributions to complex censored survival data using GLIM. *Appl. Stat.*, **29**, 156–163

Anderson, M.W., Hoel, D.G. & Kaplan, N.L. (1980) A general scheme for the incorporation of pharmacokinetics in low-dose risk estimation for chemical carcinogenesis: Example – vinyl chloride. *Toxicol. appl. Pharmacol.*, **55**, 154–161

Andervont, H.B. (1944) Influence of environment on mammary cancer in mice. *J. natl Cancer Inst.*, **4**, 579–581

Anscombe, F.J. (1949) The statistical analysis of insect counts based on the negative binomial distribution. *Biometrics*, **5**, 165–173

Anscombe, F.J. (1950) Sampling theory of the negative binomial and logarithmic series distributions. *Biometrika*, **37**, 358–382

Armitage, P. (1955) Tests for linear trends in proportions and frequencies. *Biometrics*, **11**, 375–386

Armitage, P. (1966) The chi-square test for heterogeneity of proportions, after adjustment for stratification. *J. R. stat. Soc. B*, **28**, 150–163. Addendum, **29**, (1967), 197

Armitage, P. (1971) *Statistical Methods in Medical Research*, 2nd ed., New York, John Wiley & Sons

Armitage, P. (1982) The assessment of low-dose carcinogenicity. *Biometrics*, **38**, (Supplement on Current Topics in Biostatistics and Epidemiology), 119–129

Armitage, P. & Doll, R. (1954) The age distribution of cancer and a multi-stage theory of carcinogenesis. *Br. J. Cancer*, **8**, 1–12

Armitage, P. & Doll, R. (1961) *Stochastic models for carcinogenesis*. In: Neyman, J., ed., *Proceedings of the Fourth Berkeley Symposium VI*, Berkeley, CA, Univ. California Press, pp. 19–38

Armsen, P. (1955) Tables for significance tests of 2×2 contingency tables. *Biometrika*, **42**, 494–511

Arnold, D.L., Moodie, C.A., Grice, M.C., Charbonneau, S.M., Stavric, B., Collins, B.T., McGuire, P.F., Zawidzka, Z.Z. & Munro, I.C. (1980) Long-term toxicity of ortho-toluenesulfonamide and sodium-saccharin in the rat. *Toxicol. appl. Pharmacol.*, **52**, 113–152

Arnold, D.L., Krewski, D. & Munro, I.C. (1983a) Saccharin: A toxicological and historical perspective. *Toxicology*, **27**, 179–256

Arnold, D.L., Krewski, D.R., Junkins, D.B., McGuire, P.F., Moodie, C.A. & Munro, I.C. (1983b) Reversibility of ethylenethiourea-induced thyroid lesions. *Toxicol. appl. Pharmacol.*, **67**, 264–273

Arnold, D.L., Moodie, C.A., Charbonneau, S.M., Grice, H.C., McGuire, P.F., Bryce, F.R., Collins, B.T., Zawidzka, Z.Z., Krewski, D.R., Nera, E.A. & Munro, I.C. (1985) Long term toxicity of hexachlorobenzene in the rat and the effect of dietary vitamin A. *Food Chem. Toxicol.*, **23**, 779–793.

Ayer, M., Brunk, H.D., Ewing, G.M., Reid, W.T. & Silverman, E. (1955) An empirical distribution function for sampling with incomplete information. *Ann. Math. Stat.*, **26**, 641–647

Baker, R.J. & Nelder, J.A. (1978) *General Linear Interactive Modelling (GLIM) Release 3*, Oxford, Numerical Algorithms Group

Ball, J.K. (1970) Immunosuppression and carcinogenesis: Contrasting effects with 7, 12-dimethylbenz(a)anthracene, benz(a)pyrene and 3-methylcholanthrene. *J. natl Cancer Inst.*, **44**, 1–10

Barlow, R.E., Bartholomew, D.J., Bremner, J.M. & Brunk, H.D. (1972) *Statistical Inference Under Order Restrictions: The Theory and Application of Isotonic Regression*, New York, John Wiley & Sons

Bayer, L. & Cox, C. (1979) Algorithm AS 142. Exact tests of significance in binary regression models. *Appl. Stat.*, **28**, 319–324

Bergman, S.W. & Turnbull, B.W. (1983) Efficient sequential designs for destructive life testing with application to animal serial sacrifice experiments. *Biometrika*, **70**, 305–314

Berlin, B., Brodsky, J. & Clifford, P. (1979) Testing dependence in survival experiments with serial sacrifice. *J. Am. stat. Assoc.*, **74**, 5–14

Bernstein, L., Anderson, J. & Pike, M.C. (1981) Estimation of the proportional hazard in two-treatment-group clinical trials. *Biometrics*, **37**, 513–519

Berry, G. & Wagner, J.C. (1969) The application of a mathematical model describing the times of occurrence of mesotheliomas in rats following inoculation with asbestos. *Br. J. Cancer*, **23**, 582–586

Bhapkar, V.P. & Patterson, K.W. (1977) On some nonparametric tests for profile analysis of several multivariate samples. *J. Multivariate Anal.*, **7**, 265–277

Bickis, M. & Krewski, D. (1985) *The statistical design and analysis of the long-term carcinogenicity bioassay.* In: Clayson, D., Krewski, D. & Munro, I., eds, *Toxicological Risk Assessment*, Vol. I, Boca Raton, FL, CRC Press, pp. 125–147

Bishop, Y.M.M., Fienberg, S.E. & Holland, P.W. (1974) *Discrete Multivariate Analysis*, Cambridge, MA, MIT Press

Bliss, C.I. (1953) Fitting the negative binomial distribution to biological data. *Biometrics*, **9**, 176–196

Boik, R.J. (1981) A priori tests in repeated measures designs – Effects of non-sphericity. *Psychometrika,* **46,** 241–255

Box, G.E.P. (1950) Problems in the analysis of growth and wear curves. *Biometrics,* **6,** 362–389

Box, G.E.P. & Cox, D.R. (1964) An analysis of transformations. *J. R. stat. Soc. B,* **26,** 211–252

Breslow, N. (1970) A generalized Kruskal-Wallis test for comparing K samples subject to unequal patterns of censorship. *Biometrika,* **57,** 579–594

Breslow, N.E. (1974) Covariance analysis of censored survival data. *Biometrics,* **30,** 89–99

Breslow, N.E. (1975) Analysis of survival data under the proportional hazards model. *Int. stat. Rev.,* **43,** 45–57

Breslow, N.E. & Day, N.E. (1980) *Statistical Methods in Cancer Research,* Vol. 1, *The Analysis of Case-Control Studies (IARC Scientific Publications* No. 32), Lyon, International Agency for Research on Cancer

Breslow, N.E., Day, N.E., Tomatis, L. & Turusov, V.S. (1974) Associations between tumor types in a large-scale carcinogenesis study of CF-1 mice. *J. natl Cancer Inst.,* **52,** 233–239

Breslow, N.E., Edler, L. & Berger, J. (1984) A two-sample censored-data rank test for acceleration. *Biometrics,* **40,** 1049–1062

Brown, C.C. & Fears, T.R. (1981) Exact significance levels for multiple binomial testing with application to carcinogenicity screens. *Biometrics,* **37,** 763–774

Brownlee, K.A. (1965) *Statistical Theory and Methodology in Science and Engineering,* New York, John Wiley & Sons

Bryan, W.R. & Shimkin, M.B. (1943) Quantitative analysis of dose-response data obtained with three carcinogenic hydrocarbons in strain C3H male mice. *J. natl Cancer Inst.,* **3,** 503–531

Cairns, T. (1980) The ED_{01} study: Introduction, objectives and experimental design. *J. environ. Pathol. Toxicol.,* **3,** 1–7

Carlborg, F.W. (1981) Multi-stage dose-response models in carcinogenesis. *Food Cosmet. Toxicol.,* **19,** 361–365

Chakravorti, S.R. (1974) On some tests of growth curve model under Behrens-Fisher situation. *J. Multivariate Anal.,* **4,** 31–51

Chambers, E.A. & Cox, D.R. (1967) Discrimination between alternative binary response models. *Biometrika,* **54,** 573–578

Chand, N. & Hoel, D.G. (1974) *A comparison of models for determining safe levels of environmental agents.* In: Proschan, F. & Serfling, R.J., eds., *Reliability and Biometry,* Philadelphia, PA, Society of Industrial and Applied Mathematics, pp. 681–700

Chapman, D.G. & Nam, J. (1968) Asymptotic power of chi square tests for linear trends in proportions. *Biometrics,* **24,** 315–327, errata, **25** (1969), 777

Chen, H.J. (1984) Sample size determinations when two binomial proportions are very small. *Commun. Stat.-Theor. Meth.,* **A13,** 2707–2712

Chernoff, H. (1972) *Sequential Analysis and Optimal Design,* Philadelphia, PA, Society of Industrial and Applied Mathematics

Chu, I., Villeneuve, D.C., Velli, V.E., Secours, V.E. & Becking, G.C. (1981) Chronic toxicity of photomirex in the rat. *Toxicol. appl. Pharmacol.*, **59**, 268–278

Clayson, D.B. (1981) Carcinogens and carcinogenesis enhancers. *Mutat. Res.*, **86**, 217–229

Clayson, D.B., Krewski, D.R. & Munro, I.C. (1983) The power and interpretation of the carcinogenicity bioassay. *Regul. Toxicol. Pharmacol.*, **3**, 329–348

Cochran, W.G. (1943) Analysis of variance for percentages based on unequal numbers. *J. Am. stat. Assoc.*, **38**, 287–301

Cochran, W.G. (1954) Some methods for strengthening the common χ^2 tests. *Biometrics*, **10**, 417–451

Cochran, W.G. (1974) *The vital role of randomization in experiments and surveys*. In: Neyman, J., ed., *The Heritage of Copernicus: Theories Pleasing to the Mind*, Cambridge, MA, MIT Press, pp. 445–463

Conover, W.J. (1980) *Practical Nonparametric Statistics*, 2nd Ed. New York, John Wiley & Sons

Cook, P.J., Doll, R. & Fellingham, S.A. (1969) A mathematical model for the age distribution of cancer in man. *Int. J. Cancer*, **4**, 93–112

Cornfield, J. (1956) *A statistical problem arising from retrospective studies*. In: Neyman, J., ed., *Proceedings of the Third Berkeley Symposium IV*, Berkeley, CA, Univ. California Press, pp. 135–148

Cornfield, J. (1977) Carcinogenic risk assessment. *Science*, **198**, 693–699

Cox, D.R. (1958) The regression analysis of binary sequences. *J. R. stat. Soc. B*, **20**, 215–242

Cox, D.R. (1959) The analysis of exponentially distributed life-times with two types of failure. *J. R. stat. Soc. B*, **21**, 411–421

Cox, D.R. (1966) *Some procedures connected with the logistic qualitative response curve*. In: David, F.N., ed., *Research Papers in Statistics, Festschrift for J. Neyman*, New York, John Wiley & Sons, pp. 55–71

Cox, D.R. (1970) *The Analysis of Binary Data*, London, Chapman & Hall

Cox, D.R. (1972) Regression models and life-tables (with discussion). *J. R. stat. Soc. B*, **34**, 187–220

Cox, D.R. & Hinkley, D.V. (1974) *Theoretical Statistics*, London, Chapman & Hall

Cox, D.R. & Oakes, D. (1984) *Analysis of Survival Data*, London, Chapman & Hall

Cranmer, M.F., Lawrence, L.R., Konvincka, A.J. & Herrick, S.S. (1978) NCTR computer systems designed for toxicologic experimentation. I. Overview. *J. environ. Pathol. Toxicol.*, **1**, 701–709

Crouch, E. & Wilson, R. (1981) Regulation of carcinogens. *Risk Anal.*, **1**, 47–57

Crump, K.S. (1979) Dose response problems in carcinogenesis. *Biometrics*, **35**, 157–167

Crump, K.S. (1982) Designs for discriminating between binary dose response models with applications to animal carcinogenicity experiments. *Commun. Stat.-Theor. Meth.*, **A11**, 379–394

Crump, K.S. (1983) Ranking carcinogens for regulation (Letter). *Science*, **219**, 236–237

Crump, K.S. (1984) An improved procedure for low dose carcinogenic risk assessment from animal data. *J. environ. Pathol. Toxicol. Oncol.*, **5**, 339–348

Crump, K.S. & Howe, R.B. (1980) *A Small-sample Study of Permutation Tests for Detecting Teratogenic Effects (Technical Report)*, Washington DC, Ebon Research Systems

Crump, K.S. & Howe, R.B. (1985) *A review of methods for calculating confidence limits in low dose extrapolation.* In: Clayson, D., Krewski, D. & Munro, I., eds, *Toxicological Risk Assessment,* Vol. I, Boca Raton, FL, CRC Press, pp. 187–203

Crump, K.S., Hoel, D.G., Langley, C.H. & Peto, R. (1976) Fundamental carcinogenic processes and their implications for low dose risk assessment. *Cancer Res.*, **36**, 2973–2979

Crump, K.S., Guess, H.A. & Deal, K.L. (1977) Confidence intervals and test of hypotheses concerning dose response relations inferred from animal carcinogenicity data. *Biometrics*, **33**, 437–451

Cuzick, J. (1982) The efficiency of the proportions test and the logrank test for censored survival data. *Biometrics*, **38**, 1033–1039

Cuzick, J. (1985) Wilcoxon–type test for trend. *Stat. Med.*, **4**, 87–90

Daniel, D.L. (1983) *The analysis of body weight data.* Presented at the East Kent Local Group of the Royal Statistical Society, 17 March 1983

Daniel, C. & Wood, F.S. (1971) *Fitting Equations to Data: Computer Analysis of Multifactor Data for Scientists and Engineers,* New York, John Wiley & Sons

Davies, R.F., Lee, P.N. & Rothwell, K. (1974) A study of the dose response of mouse skin to cigarette smoke condensate. *Br. J. Cancer*, **30**, 146–156

Davis, R.K., Stevenson, G.T. & Busch, K.A. (1956) Tumor incidence in normal Sprague-Dawley female rats. *Cancer Res.*, **16**, 194–197

Day, N.E. & Brown, C.C. (1980) Multistage models and primary prevention of cancer. *J. natl Cancer Inst.*, **64**, 977–989

Day, T.D. (1967) Carcinogenic action of cigarette smoke condensate on mouse skin. *Br. J. Cancer*, **21**, 56–81

Dempster, A.P., Selwyn, M.R. & Weeks, B.J. (1983) Combining historical and randomized controls for assessing trends in proportions. *J. Am. stat. Assoc.*, **78**, 221–227

Dewanji, A. & Kalbfleisch, J.D. (1986) Non-parametric methods for survival/sacrifice experiments. *Biometrics* (in press)

Dinse, G.E. (1985) Testing for a trend in tumor prevalence rates: I. Nonlethal tumors. *Biometrics*, **41**, 751–770

Dinse, G.E. & Lagakos, S.W. (1982) Nonparametric estimation of lifetime and disease onset distributions from incomplete observations. *Biometrics*, **38**, 921–932

Dinse, G.E. & Lagakos, S.W. (1983) Regression analysis of tumour prevalence data. *Appl. Stat.*, **32**, 236–248, addendum, **33** (1984), 79–80

Dobson, A.J. & Gebski, V. (1986) Sample sizes for comparing two independent proportions using the continuity-corrected arc sine transformation. *Statistician*, **35**, 51–53

Doll, R. (1971) The age distribution of cancer: Implications for models of carcinogenesis. *J. R. stat. Soc. A*, **134**, 133–155

Doll, R. & Peto, R. (1978) Cigarette smoking and bronchial carcinoma: Dose and time relationships among regular smokers and life long non-smokers. *J. Epidemiol. Community Health*, **32**, 303–313

Drinkwater, N.R. & Klotz, J.H. (1981) Statistical methods for the analysis of tumor multiplicity data. *Cancer Res.*, **41**, 113–119

Druckrey, H. (1967) *Quantitative aspects in chemical carcinogenesis.* In: Truhaut, R., ed., *Potential Carcinogenic Hazards from Drugs. Evaluation of Risks (UICC Monograph Series)*, Berlin (West), Springer-Verlag, pp. 60–78

Druckrey, H., Preussmann, R., Ivankovic, S. & Schmähl, D. (1967) Organotrope carcinogene Wirkungen bei 65 verschiedenen *N*-Nitroso-Verbindungen an BD-Ratten. *Z. Krebsforsch.*, **69**, 103–201

Dunn, O.J. (1964) Multiple comparisons using rank sums. *Technometrics*, **6**, 241–252

Dunn, O.J. (1974) On multiple tests and confidence intervals. *Commun. Stat.*, **3**, 101–103

Dunnett, C.W. (1955) A multiple comparisons procedure for comparing several treatments with a control. *J. Am. stat. Assoc.*, **50**, 1096–1121

Dunnett, C.W. (1964) New tables for multiple comparisons with a control. *Biometrics*, **20**, 482–491

Dutt, J.E., Mattes, K.D. & Tao, L.C. (1975) *Tables of the Trivariate t for Comparing Three Treatments to a Control with Unequal Sample Sizes*, G.D. Searle and Co., Mathematical and Statistical Services, TR3

Dutt, J.E., Mattes, K.D., Soms, A.P. & Tao, L.C. (1976) An approximation to the maximum modulus of the trivariate T with a comparison to the exact values. *Biometrics*, **32**, 465–469

Edgington, E.S. (1980) *Randomization Analysis*, New York, Marcel Dekker

Elashoff, J.D. (1981) Repeated-measures bioassay with correlated errors and heterogeneous variances: A Monte-Carlo study. *Biometrics*, **37**, 475–482

Elashoff, R.M. & Beal, S. (1976) Two-stage screening designs applied to chemical-screening problems with binary data. *Ann. Rev. Biophys. Bioeng.*, **5**, 561–587

Elashoff, R.M. & Preston, D.L. (1977) *An investigation of the properties of some two sample two-stage designs.* In: Krishnaiah, P.R., ed., *Applications of Statistics*, Amsterdam, North-Holland Press, pp. 407–432

Elashoff, R.M., Preston, D.L. & Fears, T.R. (1979) Comparison and evaluation of some experimental designs for use in carcinogen screening. *J. natl Cancer Inst.*, **62**, 1209–1219

Environmental Protection Agency (1979) Water quality criteria: Availability. *Fed. Reg.*, **44**, 56627–56657

Faccini, I.M. & Naylor, D. (1979) Computer analysis and integration of animal pathology data. *Arch. Toxicol., Suppl. 2*, 517–520

Fare, G. (1965) The influence of number of mice in a box on experimental skin tumour production. *Br. J. Cancer*, **19**, 871–877

Fears, T.R. & Douglas, J.F. (1977a) Suggested procedure for reducing the pathology workload in a carcinogen bioassay program. Part I. *J. environ. Pathol. Toxicol.*, **1**, 125–137

Fears, T.R. & Douglas, J.F. (1977b) Suggested procedure for reducing the pathology workload in a carcinogen bioassay program. Part II. Incorporating blind pathology techniques and analysis for animals with tumors. *J. environ. Pathol. Toxicol.*, **1,** 211–222

Fears, T.R. & Schneiderman, M.A. (1974) Pathologic evaluation and the blind technique (Letter to the Editor). *Science,* **183,** 1144–1145

Fears, T.R. & Tarone, R.E. (1977) Response to "Use of statistics when examining lifetime studies in rodents to detect carcinogenicity." *J. Toxicol. environ. Health,* **3,** 629–632

Fears, T.R., Tarone, R.E. & Chu, K.C. (1977) False-positive and false-negative rates for carcinogenicity screens. *Cancer Res.,* **37,** 1941–1945

Felsky, G., Villeneuve, D.C. & Farmer, D. (1979) An interactive toxicological data handling system for a PDP-12 computer. *Comput. Programs Biomed.,* **10,** 75–80

Festing, M.F.W. (1979) *Inbred Strains in Biomedical Research,* Oxford, Oxford University Press

Fisher, J.C. & Hollomon, J.H. (1951) A hypothesis for the origin of cancer foci. *Cancer,* **4,** 916–918

Fisher, R.A. (1935) The logic of inductive inference. *J. R. stat. Soc.,* **98,** 39–54

Food Safety Council (1978) Proposed system for food safety assessment. Chronic toxicity testing. *Food Cosmet. Toxicol.,* **16** (*Suppl. 2*), 97–108

Fox, J.G., Thibert, P., Arnold, D.L., Krewski, D.R. & Grice, H.C. (1979) Toxicology studies. II. The laboratory animal. *Food Cosmet. Toxicol.,* **17,** 661–675

Frangos, C.C. & Stone, M. (1984) On jackknife, cross-validity and classical methods of estimating a proportion with batches of different sizes. *Biometrika,* **71,** 361–366

Freundt, K.J. (1982) Mixed exposures to chemical hazards. *Occup. Health Saf.,* **51,** 10–13, 39–42

Frith, C.H., Herrick, S.S. & Konvicka, A.J. (1977) Computer assisted collection and analysis of pathology data. *J. natl Cancer Inst.,* **58,** 1717–1727

Frith, C.H., Baetcke, K.P., Nelson, C.J. & Schieferstein, G. (1979) Importance of the mouse liver tumor in carcinogenesis bioassay studies using benzidine dihydrochloride as a model. *Toxicol. Lett.,* **4,** 507–518

Frith, C.H., Boothe, A.D., Greenman, D.L. & Farmer, J.H. (1980) Correlations between gross and microscopic lesions in carcinogenic studies in mice. *J. environ. Pathol. Toxicol.,* **3,** 139–153

Gail, M.H., Santner, T.J. & Brown, C.C. (1980) An analysis of comparative carcinogenesis experiments based on multiple times to tumor. *Biometrics,* **36,** 255–266

Gainer, J.H. & Pry, T.W. (1972) Effects of arsenicals on viral infections in mice. *Am. J. vet. Res.,* **33,** 2299–2307

Gart, J.J. (1962) On the combination of relative risks. *Biometrics,* **18,** 601–610

Gart, J.J. (1970) Point and interval estimation of the common odds ratio in the combination of 2 × 2 tables with fixed marginals. *Biometrika,* **57,** 471–475

Gart, J.J. (1972) Contribution to the discussion on the paper by D.R. Cox, Regression Models and Life Tables. *J. R. stat. Soc. B,* **34,** 212–213

Gart, J.J. (1975) Letter to the Editor. *Br. J. Cancer,* **31,** 696–697

Gart, J.J. (1976) *Statistical analysis of the first mouse skin painting study.* In: Gori, G.B., ed., *Toward Less Hazardous Cigarettes Report No. 1 (DHEW Publication No.*

(*NIH*)*76–905*), Washington DC, Department of Health, Education, and Welfare, pp. 109–121

Gart, J.J. (1977) Exact and approximate tests for relative potency. *Bull. int. stat. Inst.*, **47**, 172–175

Gart, J.J. & Thomas, D.G. (1972) Numerical results on approximate confidence limits for the odds ratio. *J. R. stat. Soc. B*, **34**, 441–447

Gart, J.J., Chu, K.C. & Tarone, R.E. (1979) Statistical issues in interpretation of chronic bioassay tests for carcinogenicity. *J. natl Cancer Inst.*, **62**, 957–974

Gaylor, D.W., Chen, J.J. & Kodell, R.L. (1985a) Experimental designs of bioassays due for screening and low dose extrapolation. *Risk Anal.*, **5**, 9–16

Gaylor, D.W., Chen, J.J., Greenman, D.L. & Thompson, C.H. (1985b) Occurrence of tumors among litters of BALB/C female mice. *J. natl Cancer Inst.*, **74**, 803–809

Gehring, P.J. & Blau, G.E. (1977) Mechanisms of carcinogenesis: dose response. *J. environ. Pathol. Toxicol.*, **1**, 163–179

Gehring, P.J., Watanabe, P.G. & Park, C.N. (1978) Resolution of dose response toxicity data for chemicals requiring metabolic activation: Example – vinyl chloride. *Toxicol. appl. Pharmacol.*, **44**, 581–591

Geisser, S. (1980) *Growth curve analysis.* In: Krishnaiah, P., ed., *Handbook of Statistics*, Vol. 1, Amsterdam, North-Holland, pp. 89–115

Gladen, B. (1979) The use of the jackknife to estimate proportions from toxicological data in the presence of litter effects. *J. Am. stat. Assoc.*, **74**, 278–283

Glasbey, C.A. (1979) Correlated residuals in non-linear regression applied to growth data. *Appl. Stat.*, **28**, 251–259

Gold, L.S., Sawyer, C.B., Magaw, R., Backman, G.M., de Veciana, M., Levinson, R., Hooper, N.K., Havender, W.R., Bernstein, L., Peto, R., Pike, M.C. & Ames, B.N. (1984) A carcinogenic potency database of the standardized results of animal bioassays. *Environ. Health Perspect.*, **58**, 9–319

Goodman, D.G., Ward, J.M., Squire, R.A., Chu, K.C. & Linhart, M.S. (1979) Neoplastic and nonneoplastic lesions in aging F344 rats. *Toxicol. appl. Pharmacol.*, **48**, 237–248

Graham, S.L., Davies, K.J., Hansen, W.H. & Graham, C.H. (1975) Effects of prolonged ethylene thiourea ingestion on the thyroid of the rat. *Food Cosmet. Toxicol.*, **13**, 493–499

Greenman, D.L., Kodell, R.L. & Sheldon, W.G. (1984) Association between cage shelf level and spontaneous and induced neoplasms in mice. *J. natl Cancer Inst.*, **73**, 107–113

Grice, H.C., Munro, I.C., Krewski, D.R. & Blumenthal, H. (1981) In utero exposure in chronic toxicity/carcinogenicity studies. *Food Cosmet. Toxicol.*, **19**, 373–379

Grizzle, J.E. & Allen, D.M. (1969) Analysis of growth and dose response curves. *Biometrics*, **25**, 357–381

Guess, H.A. & Crump, K.S. (1978) Maximum likelihood estimation of dose response functions subject to absolutely monotonic constraints. *Ann. Stat.*, **6**, 101–111

Hammond, E.C., Selikoff, I.J. & Seidman, H. (1979) Asbestos exposure, cigarette smoking and death rates. *Ann. N.Y. Acad. Sci.*, **330**, 473–490

Hartley, H.O. & Sielken, R.L. Jr (1977) Estimation of "safe doses" in carcinogenic experiments. *Biometrics,* **33,** 1–30

Hartley, H.O. & Sielken, R.L. (1978) *Development of Statistical Methodology for Risk Estimation: Final Report,* College Station, TX, Texas A & M University, Institute of Statistics

Haseman, J.K. (1977) Response to "Use of statistics when examining life time studies in rodents to detect carcinogenicity". *J. Toxicol. environ. Health,* **3,** 633–636

Haseman, J.K. (1978) Exact sample sizes for use with the Fisher–Irwin test for 2×2 tables. *Biometrics,* **34,** 106–109

Haseman, J.K. (1983a) Patterns of tumor incidence in two-year cancer bioassay feeding studies in Fischer 344 rats. *Fund. appl. Toxicol.,* **3,** 1–9

Haseman, J.K. (1983b) A re-examination of false-positive rates for carcinogenesis studies. *Fund. appl. Toxicol.,* **3,** 334–339

Haseman, J.K. (1985) *False positive issues in carcinogenicity testing: An examination of 16 studies with dual control groups.* In: *Proceedings of the Symposium on Long-Term Animal Carcinogenicity Studies: A Statistical Perspective,* Washington DC, American Statistical Association, pp. 73–80

Haseman, J.K. & Hoel, D.G. (1979) Statistical design of toxicity assays: role of genetic structure of test animal population. *J. Toxicol. environ. Health,* **5,** 89–101

Haseman, J.K. & Hogan, M.D. (1975) Selection of the experimental unit in teratology studies. *Teratology,* **12,** 165–172

Haseman, J.K. & Kupper, L.L. (1979) Analysis of dichotomous data from certain toxicology experiments. *Biometrics,* **35,** 281–293

Haseman, J.K. & Soares, E.R. (1976) The distribution of fetal death in control mice and its implications on statistical tests for dominant lethal effects. *Mutat. Res.,* **41,** 277–288

Hatch, A.M., Winberg, G.S., Zawidzka, Z., Cann, M., Airth, J.M. & Grice, H.C. (1965) Isolation syndrome in the rat. *Toxicol. appl. Pharmacol.,* **7,** 737–745

Haybittle, J.L. & Freedman, L.S. (1979) Some comments on the logrank test statistic in clinical trial applications. *Statistician,* **28,** 199–208

Haux, R. (1985) Analysis of profiles based on ordinal classification functions and rank tests. *Biom. J.,* **27,** 607–622

Health and Welfare Canada (1975) *The Testing of Chemicals for Carcinogenicity, Mutagenicity and Teratogenicity,* Ottawa

Healy, M.J.R. (1972) Animal litters as experimental units. *Appl. Stat.,* **21,** 155–159

Hennings, H. (1986) *Tumor promotion and progression in mouse skin.* In: Barret, J.C., ed., *Mechanisms of Environmental Carcinogenesis,* Vol. 2, Boca Raton, FL, CRC Press (in press)

Herrick, S.S., Davis, C., Donnelly, D.V., Lockhart, T., Marek, L. & Russel, H. (1983) Histopathology automated system. *Drug Inf. J.,* **17,** 287–295

Hewlett, P.S. & Plackett, R.L. (1979) *An Introduction to the Interpretation of Quantal Responses in Biology,* London, Arnold

Hitchcock, S.E. (1966) Tests of hypotheses about the parameters of the logistic function. *Biometrika,* **53,** 535–544

Hoel, D.G. (1980) Incorporation of background in dose-response models. *Fed. Proc.,* **39,** 73–75

Hoel, D.G. (1983) *Conditional two-sample tests with historical controls.* In: Sen, P.K., ed., *Contributions to Statistics: Essays in Honour of Norman L. Johnson,* Amsterdam, North-Holland, pp. 229–236

Hoel, D.G. & Walburg, H.E. (1972) Statistical analysis of survival experiments. *J. natl Cancer Inst.,* **49,** 361–372

Hoel, D.G. & Yanagawa, T. (1986) Incorporating historical controls in testing for trends in proportions. *J. Am. stat. Assoc.* (in press)

Hoel, D.G., Kaplan, N.L. & Anderson, M.W. (1983) Implication of nonlinear kinetics on risk estimation in carcinogenesis. *Science,* **219,** 1032–1037

Hollander, M. & Wolfe, D.A. (1973) *Nonparametric Statistical Methods,* New York, John Wiley & Sons

Hoover, K.L., Ward, J.M. & Stinson, S.F. (1980) Histopathologic differences between liver tumors in untreated (C57BL/6 × C3H) F1 (B6C3F1) mice and nitrofen-fed mice. *J. natl Cancer Inst.,* **65,** 937–948

Howell, S.B., Dean, J.H. & Law, L.W. (1975) Defects in cell-mediated immunity during growth of a syngeneic simian virus-induced tumor. *Int. J. Cancer,* **15,** 152–169

Huber, P.J. (1981) *Robust Statistics,* New York, John Wiley & Sons

Hueper, W.C. & Wolfe, H.D. (1937) Experimental production of aniline tumors of the bladder in dogs. *Am. J. Pathol.,* **13,** 656

Hueper, W.C., Wiley, F.H. & Wolfe, H.D. (1938) Experimental production of bladder tumors in dogs by administration of beta-naphthylamine. *J. ind. Hyg.,* **20,** 46–84

Hulse, E.V., Mole, R.H. & Papworth, D.G. (1968) Radiosensitivities of cells from which radiation-induced skin tumours are derived. *Int. J. Radiat. Biol.,* **14,** 437–444

IARC (1980) *IARC Monographs on the Evaluation of the Carcinogenic Risk of Chemicals to Humans, Supplement 2, Long-term and Short-term Screening Assays for Carcinogens: A Critical Appraisal,* Lyon, International Agency for Research on Cancer

IARC (1982a) *Information Bulletin on the Survey of Chemicals being Tested for Carcinogenicity,* No. 10, Lyon, International Agency for Research on Cacer

IARC (1982b) *IARC Monographs on the Evaluation of the Carcinogenic Risk of Chemicals to Humans, Supplement 4, Chemicals, Industrial Processes and Industries Associated with Cancer in Humans. IARC Monographs, Volumes 1 to 29,* Lyon, International Agency for Research on Cancer.

Iball, J. (1939) The relative, potency of carcinogenic compounds. *Am. J. Cancer,* **35,** 188–190

Interagency Regulatory Liaison Group (1979) Scientific bases for identification of potential carcinogens and estimation of risks. *J. natl Cancer Inst.,* **63,** 241–268

Interdisciplinary Panel on Carcinogenicity (1984) Criteria for evidence of chemical carcinogenicity. *Science,* **225,** 682–687

International Life Sciences Institute (1984a) *Age-associated (geriatric) pathology: Its impact on long-term toxicity studies.* In: Grice, H.C. ed., *Current Issues in Toxicology,* New York, Springer-Verlag, pp. 50–107

International Life Sciences Institute (1984b) *The selection of doses in chronic toxicity/carcinogenicity studies.* In: Grice, M.C., ed., *Current Issues in Toxicology,* New York, Springer-Verlag, pp. 6–49

Irwin, J.O. & Goodman, N. (1946) The statistical treatment of measurements of the carcinogenic properties of tars (Part I) and mineral oils (Part II). *J. Hyg,.* **44,** 362–420

Ivankovic, S. (1973) *Experimental prenatal carcinogenesis.* In: Tomatis, L. & Mohr, U., eds., *Transplacental Carcinogenesis (IARC Scientific Publications No. 4)*, Lyon, International Agency for Research on Cancer, pp. 92–99

Iverson, S. & Arley, N. (1950) On the mechanism of experimental carcinogenesis. *Acta pathol. microbiol. Scand.,* **27,** 773–803

Jonckheere, A.R. (1954) A distribution-free k-sample test against ordered alternatives. *Biometrika,* **41,** 133–145

Jones, C.A., Marlino, P.H.J., Lijinsky, W. & Huberman, E. (1981) The relationship between the carcinogenicity and mutagenicity of nitrosoamines in a hepatocyte-mediated mutagenicity assay. *Carcinogenesis,* **2,** 1075–1077

Kalbfleisch, J.D. & Prentice, R.L. (1980) *The Statistical Analysis of Failure Time Data,* New York, John Wiley & Sons

Kalbfleisch, J.D., Lawless, J.L. & MacKay, R.J. (1982) *The estimation of small probabilities and risk assessment.* In: Lind, N.C., ed., *Technological Risk,* Waterloo, University of Waterloo Press, pp. 17–26

Kalbfleisch, J.D., Krewski, D.R. & Van Ryzin, J. (1983) Dose-response models for time-to-response toxicity data (with discussion). *Can. J. Stat.,* **11,** 25–49

Kaplan, E.L. & Meier, P. (1958) Nonparametric estimation from incomplete observations. *J. Am. stat. Assoc.,* **53,** 457–481

Karlson, P. (1965). *Introduction to Modern Biochemistry,* 2nd ed., New York, Academic Press

Kempthorne, O. (1977) Why randomize? *J. Stat. Planning Inference,* **1,** 1–25

Kendall, M.G.K. & Stuart, A. (1961) *The Advanced Theory of Statistics,* Vol. 2. London, Griffin

Khatri, C.G. (1966) A note on a MANOVA model applies to problems in growth curve. *Ann. Inst. Stat. Math.,* **18,** 75–86

Kleinbaum, D.G. (1973) A generalization of the growth curve model which allows missing data. *J. Multivariate Anal.,* **3,** 117–124

Kleinman, J.C. (1973) Proportions with extraneous variance: single and independent samples. *J. Am. stat. Assoc.,* **68,** 46–54

Koch, G. G., Amara, I.A., Stokes, M.E. & Gillings (1980) Some views on parametric and nonparametric analysis for repeated measurements and selected bibliography. *Int. stat. Rev.,* **48,** 249–265

Kodell, R.L. & Nelson, C.J. (1980) An illness-death model for the study of the carcinogenic process using survival/sacrifice data. *Biometrics,* **36,** 267–277

Kodell, R.L., Shaw, G.W. & Johnson, A.M. (1982a) Nonparametric joint estimators for disease resistance and survival functions in survival/sacrifice experiments. *Biometrics,* **38,** 43–58

Kodell, R.L., Farmer, J.H., Gaylor, D.W., & Cameron, A.M. (1982b) Influence of cause-of-death assignment on time-to-tumor analyses in animal carcinogenesis studies. *J. natl Cancer Inst.,* **69,** 659–664

Koller, L.D. (1973) Immunosuppression produced by lead, cadmium and mercury. *Am. J. Vet. Res.,* **34,** 1457–1458

Konvicka, A.J., Robinson, O. & Weiss, K. (1978) NCTR computer systems designed for toxicological experimentation. II. Experiment start-up system. *J. environ. Pathol. Toxicol.*, **1**, 711–719

Korn, E.L. & Liu, P.Y. (1983) Interactive effects of mixtures of stimuli in life table analysis. *Biometrika*, **70**, 103–110

Kowalski, C.J. & Guire, K.E. (1974) Longitudinal data analysis. *Growth*, **38**, 131–169

Koziol, J.A., Maxwell, D.A., Fukushima, M., Colmerauer, M.E.M. & Pilch, Y.H. (1981) A distribution-free test for tumor-growth curve analyses with application to an animal tumor immunotherapy experiment. *Biometrics*, **37**, 383–390

Krewski, D. & Van Ryzin, J. (1981) *Dose response models for quantal response toxicity data.* In: Csorgo, M., Dawson, D.A., Rao, J.N.K. & Saleh, A.K., eds., *Statistics and Related Topics,* Amsterdam, North-Holland, pp. 201–231

Krewski, D., Crump, K.S., Farmer, J., Gaylor, D.W., Howe, D., Portier, C., Salsburg, D., Sielken, R.L. & Van Ryzin, J. (1983) A comparison of statistical methods for low dose extrapolation utilizing time-to-tumor data. *Fund. appl. Toxicol.*, **3**, 140–158

Krewski, D., Brennan, J. & Bickis, M. (1984a) The power of the Fisher permutation test in $2 \times k$ tables. *Commun. Stat.-Simul. Comp.*, **B13**, 433–448

Krewski, D., Kovar, J.G. & Bickis, M. (1984b) *Optimal experimental designs for low dose extrapolation. II. The case of nonzero background.* In: Chaubey, Y.P. & Dwivedi, T.P., eds, *Topics in Applied Statistics,* Montreal, Concordia University, pp. 167–191

Krewski, D., Smythe, R.T. & Burnett, R.T. (1985) *The use of historical control information in testing for trend in quantal response carcinogenicity data.* In: *Proceedings of the Symposium on Long-term Animal Carcinogenicity Studies: A Statistical Perspective,* Washington DC, American Statistical Association, pp. 56–62

Kuiper-Goodman, T., Krewski, D., Combley, H., Doran, M. & Grant, D.L. (1976) *Hexachlorobenzene-induced smooth endoplasmic reticulum in rat liver: a correlated stereologic and biochemical study.* In: *Proceedings of the Fourth International Congress for Stereology,* Washington DC, National Bureau of Standards Special Publication, Vol. 431, pp. 351–354

Kulwich, B.A., Hardisty, J.F., Gilmore, C.E. & Ward, J.M. (1980) Correlations between gross observations of tumours and neoplasms diagnosed microscopically in carcinogenesis bioassays in rats. *J. environ. Pathol. Toxicol.*, **3**, 281–287

Kupper, L.L. & Haseman, J.K. (1978) The use of a correlated binomial model for the analysis of certain toxicological experiments. *Biometrics*, **34**, 69–76

Kuschner, M., Laskin, S., Drew, R.T., Cappiello, V. & Nelson, N. (1975) Inhalation carcinogenicity of alpha halo ethers: III. Lifetime and limited period inhalation studies with bis(chloromethyl)ether at 0.1 ppm. *Arch. environ. Health*, **30**, 73–77

Lagakos, S.W. (1982) An evaluation of some two-sample tests used to analyze animal carcinogenicity experiments. *Util. Math.*, **21B**, 239–260

Lagakos, S.W. & Louis, T.A. (1985) *The statistical analysis of rodent tumorigenicity experiments.* In: Clayson, D.B., Krewski, D. & Munro, I., eds, *Toxicological Risk Assessment,* Vol. I, Boca Raton, FL, CRC Press, pp. 149–163

Lagakos, S. & Mosteller, F. (1981) A case study of statistics in the regulatory process: The FD & C Red No. 40 experiments. *J. natl Cancer Inst.*, **66**, 197–212

Lagakos, S.W. & Ryan, L.M. (1985) On the representativeness assumption in prevalence tests of carcinogenicity. *Appl. Stat.*, **34**, 54–62

Latta, R.B. (1981) A Monte Carlo study of some two-sample rank tests with censored data. *J. Am. stat. Assoc.*, **76**, 713–719

Lawless, J.F. (1984) *Some problems concerning experimental design for extrapolation.* In: Chaubey, Y.P. & Dwivedi, T.P., eds, *Topics in Applied Statistics,* Montreal, Concordia University, pp. 357–366

Lawrence, L.R., Konvicka, A.J., Ezell, R., Applegat, J., Green, G. & Fernstrom, E.B. (1979) NCTR computer systems designed for toxicologic experimentation. IV. Experiment information system. *J. environ. Pathol. Toxicol.*, **2**, 1011–1019

Lee, E.T. (1980) *Statistical Methods for Survival Data Analysis,* Belmont, Lifetime Learning Publications

Lee, P.N. (1975) *Final Analysis Experiment 1.1.1.9. The Effect of Stopping Painting (Tobacco Research Council Document M432),* London, Tobacco Research Council

Lee, P.N. (1979) *PSC7-Mouse Changeover Experiment A.1.5.6. Statistical Analysis of Visible Skin Tumour Data (Tobacco Advisory Council Document TA 51),* London, Tobacco Research Council

Lee, P.N. & O'Neill, J.A. (1971) The effect of both of time and dose applied on tumour incidence rate in benzopyrene skin painting experiments. *Br. J. Cancer,* **25**, 759–770

Lee, P.N., Rothwell, K. & Whitehead, J.K. (1977) Fractionation of mouse skin carcinogens in cigarette smoke condensate. *Br. J. Cancer,* **35**, 730–742

Leeper, J.D. & Woolson, R.F. (1982) Testing hypotheses for the growth curve model when the data are incomplete. *J. Stat. Comp. Simul. B,* **15**, 97–107

Littlefield, N.A., Farmer, J.H., Gaylor, D.W. & Sheldon, W.G. (1980a) Effects of dose and time in a long-term, low-dose carcinogenic study. *J. environ. Pathol. Toxicol.*, **3**, 17–34

Littlefield, N.A., Greenman, D.L. & Farmer, J.H. (1980b) Effect of continuous and discontinued exposure to 2-AAF on urinary bladder hyperplasia and neoplasia. *J. environ. Pathol. Toxicol.*, **3**, 35–54.

Louis, T.A. & Orav, E.J. (1985) *Adaptive sacrifice plans for the carcinogen bioassay.* In: *Proceedings of the Symposium on Long-Term Animal Carcinogenicity Studies: A Statistical Perspective,* Washington DC, American Statistical Association, pp. 36–41

Maltoni, C. (1975) The value of predictive experimental bioassays in occupational and environmental carcinogeneses. An example: vinyl chloride. *Ambio,* **4**, 18–23

Mantel, N. (1963) Chi-square tests with one degree of freedom; extensions of the Mantel-Haenszel procedure. *J. Am. stat. Assoc.*, **58**, 690–700

Mantel, N. (1966) Evaluation of survival data and two new rank order statistics arising in its consideration. *Cancer Chemother. Rep.*, **50**, 163–170

Mantel, N. (1980) Assessing laboratory evidence for neoplastic activity. *Biometrics,* **36**, 381–399

Mantel, N. (1983) Commentary on the work of Michalek and Mihalko relative to litter-matched data. *Stat. Med.*, **2**, 323–326

Mantel, N. & Bryan, W.R. (1961) "Safety" testing of carcinogenic agents. *J. natl Cancer Inst.*, **27**, 455–470

Mantel, N. & Ciminera, J.L. (1979) Use of logrank scores in the analysis of litter-matched data on time to tumor appearance. *Cancer Res., 39,* 4308–4315

Mantel, N. & Haenszel, W. (1952) Statistical aspects of the analysis of data from the retrospective studies of disease. *J. natl Cancer Inst., 22,* 719–748

Mantel, N., Bohidar, W.R. & Ciminera, J.L. (1977) Mantel-Haenszel analyses of litter-matched time-to-response data, with modification for recovery of interlitter information. *Cancer Res., 37,* 3863–3868

Marascuilo, L.A. & McSweeney, M. (1967) Nonparametric post hoc comparisons for trend. *Psychol. Bull., 67,* 401–412

McCullagh, P. (1980) Regression models for ordinal data (with discussion). *J. R. stat. Soc. B, 42,* 109–142

McKnight, B. (1981) *Testing for Differences in Tumour Incidence.* PhD thesis, Madison, WI, University of Wisconsin

McKnight, B. (1985) *Discussion of session on statistical tests for carcinogenic effects.* In: *Proceedings of the Symposium on Long-Term Animal Carcinogenicity Studies: A Statistical Perspective,* Washington DC, American Statistical Association, pp. 107–111

McKnight, B. & Crowley, J. (1984) Tests for differences in tumor incidence based on animal carcinogenesis experiments. *J. Am. stat. Assoc., 79,* 639–648

Meng, C.Y.K. (1985) *A Bayesian approach to the multiplicity problem for significance testing with binomial data.* In: *Proceedings of the Symposium on Long-Term Animal Carcinogenicity Studies: A Statistical Perspective,* Washington DC, American Statistical Association, pp. 66–72

Meselson, M. & Russel, K. (1977) *Comparisons of carcinogenic and mutagenic potency.* In: Hiatt, H.H., Watson, J.D. & Winston, J.A., eds, *Origins of Human Cancer,* Book C, Cold Spring Harbor, NY, Cold Spring Harbour Laboratory, pp. 1473–1481

Métivier, H., Wahrendorf, J. & Masse, R. (1984) Multiplicative effect of inhaled plutonium oxide and benzo(a)pyrene on lung carcinogenesis in rats. *Br. J. Cancer, 50,* 215–221

Michalek, J.E. & Mihalko, D. (1983) On the use of logrank scores in the analysis of litter-matched data on time to tumour appearance. *Stat. Med., 2,* 315–326

Michalek, J.E. & Mihalko, D. (1984) Linear rank procedures on litter-matched data. *Biometrics, 40,* 487–491

Miescher, G., Almasy, F. & Zehender, F. (1941) Besteht ein Zusammenhang zwischen dem Benzpyrengehalt und der carcinogenen Wirkung des Teers? *Schweiz. med. Wochenschr., 34,* 1002–1007

Miller, R.G., Jr (1981a) *Simultaneous Statistical Inference,* 2nd ed., Heidelberg, Springer-Verlag

Miller, R.G., Jr. (1981b) *Survival Analysis,* New York, NY, John Wiley & Sons

Miller, R.G., Jr (1986) *Beyond ANOVA, Basics of Applied Statistics,* New York, John Wiley & Sons

Mitchell, T.J. & Turnbull, B.W. (1979) Log-linear models in the analysis of disease prevalence data from survival/sacrifice experiments. *Biometrics, 35,* 221–234

Moolgavkar, S.H. & Knudson, A.G. (1981) Mutation and cancer: A model for human carcinogenesis. *J. natl Cancer Inst.*, **66**, 1037–1052

Morrison, D.F. (1976) *Multivariate Statistical Methods,* 2nd ed., New York, McGraw Hill

Munro, I.C. (1977) Considerations in chronic toxicity testing: The chemical, the dose, the design. *J. environ. Pathol. Toxicol,* **1**, 183–197

Nam, J. (1984) Approximate formula for sample size determinations for detecting a linear trend in proportions (unpublished manuscript)

National Cancer Institute (1978) *Bioassay of 1,2-Dichloroethane for Possible Carcinogenicity (Technical Report Series No. 55),* DHEW Publication No. (NIH)78–1361, Washington DC, US Department of Health, Education & Welfare

National Toxicology Program (1982) *Carcinogenesis Bioassay of 2,3,7,8-Tetrachlorodibenzo-p-dioxin (CAS No. 1746–01–6) in Osborne-Mendel Rats and B63F₁ Mice (Gavage Study) (National Toxicology Program Technical Report Series No. 209),* Bethesda, MD, US Department of Health and Human Services

Naylor, D. (1978) The computerization of histopathological data in toxicological laboratory studies using SNOP. *Meth. Inf. Med.,* **17**, 272–279

Nelson, K., Cory, J., Hellstrom, I. & Hellstrom, K.E. (1980) *T-T* Hybridoma product specifically suppresses tumor immunity. *Proc. natl Acad. Sci. USA,* **77**, 2866–2870

Nowinsky, M. (1876) Zur Frage über die Impfung der krebsigen Geschwülste. *Zentralbl. med. Wissensch.,* **14**, 790–791

Ochi, Y. & Prentice, R.L. (1984) Likelihood inference in a correlated probit regression model. *Biometrika,* **71**, 531–543

OECD Long-Term Expert Group (1981) *Test Guidelines for Carcinogenicity Studies,* Paris, Organization for Economic and Cooperative Development

Page, E.B. (1963) Ordered hypotheses for multiple treatments: A significance test for linear ranks. *J. Am. stat. Assoc.,* **58**, 216–230

Parish, S. (1981) *Exploiting Animal Tumour Data using Multistage Models.* D.Phil Thesis, Oxford, Oxford University

Parmiani, G., Colnaghi, M.I. & Della Porta, G. (1971) Immunodepression during urethane and N-nitrosomethylurea leukaemogenesis in mice. *Br. J. Cancer,* **25**, 354–364

Pendergast, J.F. & Broffitt, J.D. (1985) Robust estimation in growth curves. *Commun. Stat.-Simul. Comp.,* **B14**, 1919–1940

Peraino, C., Fry, R.J.M. & Staffeldt, E. (1977) Effects of varying the onset and duration of exposure to phenobarbital on its enhancement of 2-acetylaminofluorene-induced hepatic tumorigenesis. *Cancer Res.,* **37**, 3623–3627

Perry, L.L. & Greene, M.I. (1981) *T* cell subset interactions in the regulation of syngeneic tumor immunity. *Fed. Proc.,* **40**, 39–44

Peto, R. (1974) Guidelines on the analysis of tumour rates and death rates in experimental animals. *Br. J. Cancer,* **29**, 101–105

Peto, R. (1977) *Epidemiology, multistage models and short-term mutagenicity tests.* In: Hiatt, H.H., Watson, J.D. & Winsten, J.A., eds, *Origins of Human Cancer,* Book C, Cold Spring Harbor, NY, Cold Spring Harbor Laboratory, pp. 1403–1428

Peto, R. & Lee, P. (1973) Weibull distributions for continuous carcinogenesis experiments. *Biometrics, 29,* 457–470

Peto, R. & Peto, J. (1972) Asymptotically efficient rank invariant test procedures (with discussion). *J. R. stat. Soc. A,* **135,** 185–206

Peto, R. & Pike, M.C. (1973) Conservatism of the approximation $\sum (O-E)^2/E$ in the logrank test for survival data or tumor incidence data. *Biometrics, 29,* 579–584

Peto, R., Lee, P.N. & Paige, W.S. (1972) Statistical analysis of the bioassay of continuous carcinogens. *Br. J. Cancer,* **26,** 258–261

Peto, R., Roe, F.J.C., Lee, P.N., Levy, L. & Clack, J. (1975) Cancer and ageing in mice and men. *Br. J. Cancer,* **32,** 411–426

Peto, R., Pike, M.C., Day, N.E., Gray, R.G., Lee, P.N., Parish, S., Peto, J., Richards, S. & Wahrendorf, J. (1980) *Guidelines for simple, sensitive significance tests for carcinogenic effects in long-term animal experiments.* In: *Long-term and Short-term Screening Assays for Carcinogens: A Critical Appraisal (IARC Monographs Supplement 2),* Lyon, International Agency for Research on Cancer, pp. 311–426

Peto, R., Gray, R., Brantom, P. & Grasso, P. (1984) *Nitrosamine carcinogenesis in 5120 rodents: chronic administration of sixteen different concentrations of NDEA, MPYR and NPIP in the water of 4440 inbred rats, with parallel studies on NDEA slone of the effect of age of starting (3, 6 or 20 weeks) and of species (rats, mice or hamsters).* In: O'Neill, I.K., Von Borstel, R.C., Miller, T.C., Long, J. & Bartsch, H., eds, N-*Nitroso Compounds: Occurrence, Biological Effects and Relevance to Human Cancer (IARC Scientific Publications No. 57),* Lyon, International Agency for Research on Cancer, pp. 627–665

Pike, M.C. (1966) A method of analysis of a certain class of experiments in carcinogenesis. *Biometrics,* **22,** 142–161

Pike, M.C. (1972) Contribution to discussion on the paper by R. Peto and J. Peto. *J. R. stat. Soc. A,* **135,** 201–203

Pitot, H.C. & Sirica, A.E. (1980) The stages of initiation and promotion in hepato-carcinogenesis. *Biochem. biophys. Acta,* **605,** 191–216

Port, R., Schmähl, D. & Wahrendorf, J. (1976) Some examples of dose-response studies in chemical carcinogenesis. *Oncology,* **33,** 66–71

Portier, C. (1981) *Optimal Bioassay Design under the Armitage-Doll Multistage Model (Technical Report),* Department of Biostatistics, Chapel Hill, NC, University of North Carolina

Portier, C. & Hoel, D. (1983a) Low-dose-rate extrapolation using the multistage model. *Biometrics,* **39,** 897–906

Portier, C. & Hoel, D. (1983b) Optimal design of the chronic animal bioassay. *J. Toxicol. Environ. Health,* **12,** 1–19

Portier, C. & Hoel, D. (1984a) Design of animal carcinogenicity studies for goodness-of-fit of multistage models. *Fundam. appl. Toxicol.,* **4,** 949–959

Portier, C. & Hoel, D. (1984b) Type 1 error of trend tests in proportions and the design of cancer screen. *Commun. Stat.-Theor. Meth.,* **13,** 1–14

Potthoff, R.F. & Roy, S.N. (1964) A generalized multivariate analysis of variance model useful especially for growth curve problems. *Biometrika,* **51,** 313–326

Prentice, R.L. (1978) Linear rank tests with right censored data. *Biometrika,* **65,** 167–179

Prentice, R.L. & Marek,, P. (1979) A qualitative discrepancy between censored data rank tests. *Biometrics,* **35,** 861–867

Prestele, H., Gaus, W. & Horbach, L. (1979) A procedure for comparing groups of time-dependent measurements. *Meth. Inf. Med.,* **18,** 84–88

Purchase, I.F.H. (1980) Inter-species comparisons of carcinogenicity. *Br. J. Cancer,* **41,** 454–468

Radhakrishna, S. (1965) Combination of results from several 2×2 contingency tables. *Biometrics,* **21,** 86–98

Rai, K. & Van Ryzin, J. (1981) A generalized multihit dose-response model for low-dose extrapolation. *Biometrics,* **37,** 341–352

Rai, K. & Van Ryzin, J. (1983) *A Dose Response Model Incorporating Michaelis– Menten Kinetics (Technical Report B–29),* Division of Biostatistics, New York, Columbia University

Rao, J.N.K. & Scott, A.J. (1981) The analysis of categorical data from complex sample surveys: chi-squared tests for goodness of fit and independence in two-way tables. *J. Am. stat. Assoc.,* **76,** 221–230

Reznik, G. & Ward, J.M. (1979) Carcinogenicity of the hair-dye component 2-nitro-p-phenylenediamine: Introduction of eosinophilic hepatocellular neoplasms in female B6C3F1 mice. *Food Cosmet. Toxicol.,* **17,** 493–500

Roe, F.J.C. & Lee, P.N. (1984) *Histopathological Data Recording, Processing, Reporting and Statistical Analysis, using Computer Program ROELEE 84* (Available from P.N. Lee, 25 Cedar Road, Sutton, Surrey, SM2 5DG, UK

Rosenkranz, G. (1982) APL-programs for the analysis of carcinogenicity experiments. *Comput. Programs Biomed.,* **15,** 87–92

Rothman, K.J. & Boice, J.D., Jr (1979) *Epidemiologic Analysis with a Programmable Calculator (NIH Publication No. 79–1649),* Washington DC, US Government Printing Office

Salsburg, D.S. (1977) Use of statistics when examining lifetime studies in rodents to detect carcinogenicity. *J. Toxicol. environ. Health,* **3,** 611–628

Sandland, R.L. & McGilchrist, C.A. (1979) Stochastic growth curve analysis. *Biometrics,* **35,** 255–271

Sawyer, C., Peto, R., Bernstein, L. & Pike, M.C. (1984) Calculation of carcinogenic potency from long-term animal carcinogenesis experiments. *Biometrics,* **40,** 27–40

Schach, S. (1982) An elementary method for the statistical analysis of growth curves. *Metrika,* **29,** 271–282

Schwertman, N.C., Magrey, J.M. & Fridshal, D. (1981) On the analysis of incomplete growth curve data, a Monte-Carlo study of two nonparametric procedures. *Commun. Stat.-Simul. Comp.* **B10,** 51–66

Scientific Review Panel (1983) *Saccharin: An Update,* Duke University Medical Centre

Scribner, J.D., Scribner, N.K., McKnight, B. & Mottet, N.K. (1983) Evidence for a new model for tumour progression from carcinogenesis and tumor promotion studies with 7-bromomethylbenz[a]anthracene. *Cancer Res.,* **43,** 2034–2041

Searle, S.R. (1971) *Linear Models,* New York, John Wiley & Sons

Segreti, A.C. & Munson, A.E. (1981) Estimation of the median lethal dose when responses within a litter are correlated. *Biometrics, 37,* 153–156

Selwyn, M.R., Roth, A.J. & Weeks, B.J. (1985) *The weighted prevalence method for analyzing nonlethal tumor data.* In: *Proceedings of the symposium on Long-Term Animal Carcinogenicity Studies: A Statistical Perspective,* Washington DC, American Statistical Association, pp. 36–41

Shimkin, M.B. (1977) *Contrary to Nature* (*NIH 76–720*), Washington DC, US Department of Health, Education, and Welfare

Shimkin, M.B. & Stoner, G.D. (1975) Lung tumors in mice: Application to carcinogenesis bioassay. *Adv. Cancer Res., 21,* 2–58

Shirley, E.A.C. & Hickling, R. (1981) An evaluation of some statistical methods for analysing numbers of abnormalities found amongst litters in teratology studies. *Biometrics, 37,* 819–829

Shubik, P. (1985) *Saccharin/Cyclamates: Laboratory evidence.* In: Wald, N.J. & Doll, R., eds, *Interpretation of Negative Epidemiological Evidence for Carcinogenicity* (*IARC Scientific Publications No. 65*), Lyon, International Agency for Research on Cancer, pp. 125–128

Siemiatycki, J. & Thomas, D.C. (1981) Biological models and statistical interactions: an example from multistage carcinogenesis. *Int. J. Epidemiol., 10,* 383–387

Sigg, E.B., Day, C. & Colombo, C. (1966) Endocrine factors in isolation-induced aggressiveness in rodents. *Endocrinology, 78,* 679–684

Slud, E.V., Byar, D.P. & Green, S.B. (1984) A comparison of reflected versus test-based confidence intervals for the median survival time based on censored data. *Biometrics, 40,* 587–600

Smith, P.G., Pike, M.C., Hill, A.P., Breslow, N.E. & Day, N.E. (1981) Algorithm AS 162. Multivariate conditional logistic analysis of stratum-matched case-control studies. *Appl. Stat., 30,* 190–197

Smythe, R.T., Krewski, D. & Murdoch, D. (1986) The use of historical control information in modelling dose-response relationships in carcinogenesis. *Stat. Probab. Lett., 4,* 87–93

Snedecor, G.W. & Cochran, W.G. (1980) *Statistical Methods,* 7th ed., Ames, IA, Iowa State University Press

Society for Toxicology (1982) Animal data in hazard evaluation: Paths and pitfalls. *Fundam. appl. Toxicol., 2,* 101–107

Sokal, R.R. & Rohlf, F.J. (1981) *Biometry – The Principles and Practice of Statistics in Biological Research,* 2nd ed., San Francisco, CA, W.H. Freeman & Co.

Soms, A.P. (1977) An algorithm for the discrete Fischer's permutation test. *J. Am. stat. Assoc., 72,* 662–664

Southward, G.M. & Van Ryzin, J. (1972) *Estimating the mean of a random binomial parameter.* In: Le Cam, L.M., Neyman, J. & Scott, E.L., eds, *Proceedings of the Sixth Berkeley Symposium IV,* Berkeley, CA, Univ. California Press, pp. 249–263

Squire, R.A. (1981) Ranking animal carcinogens: A proposed regulatory approach. *Science, 214,* 877–880

Stinson, S.F., Hoover, K.L. & Ward, J.M. (1981) Quantitation of differences between spontaneous and induced liver tumors in mice with an automated image analyzer. *Cancer Lett.*, **14**, 143–150

Swenberg, J.A., Barrow, C.S., Boreiko, C.J., Heck, H. d'A., Levin, R.J., Morgan, K.T. & Starr, T.B. (1983) Non-linear biological responses to formaldehyde and their implications for carcinogenic risk assessment. *Carcinogenesis*, **4**, 945–952

Tamura, R. & Young, S.S. (1986) The incorporation of historical control information in tests of proportions: simulation study of Tarone's procedure. *Biometrics* (in press)

Tarone, R.E. (1975) Tests for trend in life table analysis. *Biometrika*, **62**, 679–682

Tarone, R.E. (1981) On the distribution of the maximum of the logrank statistic and the modified Wilcoxon statistic. *Biometrics*, **37**, 79–85

Tarone, R.E. (1982) The use of historical control information in testing for a trend in proportions. *Biometrics*, **38**, 215–220

Tarone, R.E. (1985) On heterogeneity tests based on efficient scores. *Biometrika*, **72**, 91–95

Tarone, R.E. (1986) Correcting tests for trend in proportions for skewness. *Commun. Stat.-Theor. Math.*, **A15**, 981–998

Tarone, R.E. & Gart, J.J. (1980) On the robustness of combined tests for trends in proportions. *J. Am. stat. Assoc.*, **75**, 110–116

Tarone, R.E. & Ware, J. (1977) On distribution-free tests for equality of survival distributions. *Biometrika*, **64**, 156–60

Tarone, R.E., Chu, K.C. & Ward, J.M. (1981) Variability in the rates of some common naturally occurring tumors in Fischer 344 rats and (C57BL/6M × C3H/HeN)F_1 (B6C3F_1) mice. *J. natl Cancer Inst.*, **66**, 1175–1181

Taylor, J.M. & Friedman, L. (1974) Combined chronic feeding and three-generation reproduction study of sodium saccharin in the rat. *Toxicol. appl. Pharmacol.*, **29**, Abstract 200, p. 154

Terracini, B., Magee, P.N. & Barnes, J.M. (1967) Hepatic pathology in rats on low dietary levels of dimethylnitrosamine. *Br. J. Cancer*, **21**, 559–565

Theiss, J.C. (1983) The ranking of chemicals for carcinogenic potency. *Regul. Toxicol. Pharmacol.*, **3**, 1002–1007

Thomas, D.G. (1975) Exact and asymptotic methods for the combination of 2×2 tables. *Comput. Biomed. Res.*, **8**, 423–446

Thomas, D.G., Breslow, N. & Gart, J.J. (1977) Trend and homogeneity analyses of proportions and life table data. *Comput. Biomed. Res.*, **10**, 373–381

Thomas, D.G. & Gart, J.J. (1983) Stratified trend and homogeneity analyses of proportions and life table data. *Comput. Biomed. Res.*, **16**, 116–126

Thorpe, E. & Walker, A.I.T. (1973) The toxicology of dieldrin (HEOD). II. Comparative long-term oral toxicity studies in mice with dieldrin, DDT, phenobarbitone, beta-BHC and gamma-BHC. *Food Cosmet. Toxicol.*, **11**, 433–442

Tietjen, G. (1974) Exact and approximate tests for unbalanced random effect designs. *Biometrics*, **30**, 573–581

Tolley, H.D., Burdick, D., Manton, K.G. & Stallard, E. (1978) A compartment model approach to the estimation of tumour incidence and growth: investigation of a model of cancer latency. *Biometrics*, **34**, 377–389

Tomatis, L. (1977) *The value of long-term testing for the implementation of primary prevention.* In: Hiatt, H.H., Watson, J.D. & Winsten, J.A., eds. *Origins of Human Cancer,* Book C, Cold Spring Harbor, NY, Cold Spring Harbor Laboratory, pp. 1339–1357

Tomatis, L. (1979) The predictive value of rodent carcinogenicity tests in the evaluation of human risks. *Ann. Rev. Pharmacol. Toxicol.,* **19,** 511–530

Tomatis, L., Turusov, V., Day, N. & Charles, R.T. (1972) The effect of long-term exposure to DDT on CF-1 mice. *Int. J. Cancer,* **10,** 489–506

Tomatis, L., Partensky, C. & Montesano, R. (1973) The predictive value of mouse liver tumour induction in carcinogenicity testing – A literature survey. *Int. J. Cancer,* **12,** 1–20

Tomatis, L., Turusov, V., Charles, R.T., Boiocchi, M. & Gati, E. (1974) Liver tumors in CF-1 mice exposed for limited periods to technical DDT. *Z. Krebsforsch.,* **82,** 25–35

Turnbull, B.W. & Mitchell, T.J. (1978) Exploratory analysis of disease prevalence data from survival/sacrifice experiments. *Biometrics,* **34,** 555–570

Turnbull, B.W. & Mitchell, T.J. (1984) Nonparametric estimation of the distribution of time to onset for specific diseases in survival/sacrifice experiments. *Biometrics,* **40,** 41–50

Twort, C.C. & Twort, J.M. (1930) The relative potency of carcinogenic tars and oils. *J. Hyg.,* **29,** 373–379

Twort, C.C. & Twort, J.M. (1933) Suggested methods for the standardisation of the carcinogenic activity of different agents for the skin of mice. *Am. J. Cancer,* **17,** 293–320

Ullrich, R.L. (1980) *Carcinogenesis in mice after low dose and dose rates.* In: Meyn, R.E. & Withers, H.R. eds, *Radiation Biology in Cancer Research,* New York, Raven Press, pp. 309–319

Ullrich, R.L. & Storer, J.B. (1979a) Influence of gamma irradiation on the development of neoplastic disease in mice. I. Reticular tissue tumors. *Radiat. Res.,* **80,** 303–316

Ullrich, R.L. & Storer, J.B. (1979b) Influence of gamma irradiation on the development of neoplastic disease in mice. II. Solid tumors. *Radiat. res.,* **80,** 317–324

Ullrich, R.L., Jernigan, M.C., Cosgrove, G.E., Satterfield, L.C., Bowles, N.D. & Storer, J.B. (1976) The influence of dose and dose rate on the incidence of neoplastic disease in RFM mice after neutron irradiation. *Radiat. Res.,* **68,** 115–131

Upton, G.J.G. (1982) A comparison of alternative tests for the 2×2 comparative trial. *J.R. stat. Soc. A,* **145,** 86–105

Van Ryzin, J. (1980) Quantitative risk assessment. *J. occup. Med.,* **22,** 321–326

Vos, J.G. & de Roij, T. (1972) Immunosuppressive activity of a polychlorinated biphenyl preparation on the humoral immune response in guinea pigs. *Toxicol. appl. Pharmacol.,* **21,** 549–555

Wagner, J.C., Berry, G. & Timbrell, V. (1973) Mesotheliomata in rats after inoculation with asbestos and other materials. *Br. J. Cancer,* **28,** 173–185

Wahrendorf, J. (1983) Simultaneous analysis of different tumor types in a long-term carcinogenicity study with scheduled sacrifices. *J. natl Cancer Inst.*, **70**, 915–921

Wahrendorf, J. (1984) Discussion of D.R. Cox's paper on "Interaction". *Int. stat. Rev.*, **52**, 29–30

Wahrendorf, J. & Brown, C.C. (1980) Bootstrapping a basic inequality in the analysis of joint action of two drugs. *Biometrics*, **36**, 653–657

Wahrendorf, J., Zentgraf, R. & Brown, C.C. (1981) Optimal designs for the analysis of interactive effects of two carcinogens or other toxicants. *Biometrics*, **37**, 45–54

Wahrendorf, J., Mahon, G.A.T. & Schumacher, M. (1985) A nonparametric approach to the statistical analysis of mutagenicity data. *Mutat. Res.*, **147**, 5–13

Walker, A.I.T., Thorpe, E. & Stevenson, D.E. (1973). The toxicology of dieldrin (HEOD). I. Long-term oral toxicity studies in mice. *Food Cosmet. Toxicol.*, **11**, 415–432

Walters, D.E. (1979) In defence of the arc sine approximation. *Statistician*, **28**, 219–222

Ward, J. (1984) *Pathology of toxic, preneoplastic and neoplastic lesions in rodents*. In: Douglas, J.F., ed., *Carcinogenesis – Mutagenesis*, Clifton, NJ, Humana Press, pp. 97–130

Ward, J.M. & Reznick, G. (1983) Refinements of rodent pathology and the pathologist's contribution to evaluation of carcinogenesis bioassays. *Progr. exp. Tumour Res.*, **26**, 266–291

Ward, J.M. & Vlahakis, G. (1978) Evaluation of hepatocellular neoplasms in mice. *J. natl Cancer Inst.*, **61**, 807–811

Ward, J.M. & Weisburger, E.K. (1975) Intestinal tumors in mice treated with a single injection of N-nitroso-N-butylurea. *Cancer Res.*, **35**, 1938–1943

Ward, J.M., Goodman, D.G., Griesemer, R.A., Hardisty, J.F., Schueler, R.L., Squire, R.A. & Strandberg, J.D. (1978) Quality assurance for pathology in rodents carcinogenesis tests. *J. environ. Pathol. Toxicol.*, **2**, 371–378

Ward, J.M., Goodman, D.G., Squire, R.A., Chu, K.C. & Linhart, M.S. (1979) Neoplastic and nonneoplastic lesions in aging (C57BL/6N × C3H/HeN)F1 (B6C3F$_1$) mice. *J. natl Cancer Inst.*, **63**, 849–854

Weinberger, M.A. (1973) The blind technique (Letter to the Editor). *Science*, **181**, 219–220

Weinberger, M.A. (1979) How valuable is blind evaluation in histopathologic examination in conjunction with animal studies? *Toxicol. Pathol.*, **7**, 14–17

Weindruch, R. & Walford, R.L. (1982) Dietary restriction in mice beginning at 1 year of age: effect on life-span and spontaneous cancer incidence. *Science*, **215**, 1415–1418

Weinstein, I.B. (1981) The scientific basis for carcinogen detection and primary cancer prevention. *Cancer*, **47**, 1133–1141

Weisburger, J.H. & Williams, G.M. (1981) Carcinogen testing: Current problems and new approaches. *Science*, **214**, 401–407

Whittemore, A.S. (1978) Quantitative theories of oncogenesis. *Adv. Cancer Res.*, **27**, 55–88

Whittemore, A.S. & Keller, J.B. (1978) Quantitative theories of carcinogenesis. *SIAM Rev.*, **20**, 1–30

Williams, D.A. (1975) The analysis of binary responses from toxicological experiments involving reproduction and teratogenicity. *Biometrics*, **31**, 949–952

Williams, G.M., Katayama, S. & Ohmori, T. (1981) Enhancement of hepatocarcinogenesis by sequential administration of chemicals: summation versus promotion effects. *Carcinogenesis*, **2**, 1111–1117

Wishart, J. (1938) Growth-rate determinations in nutrition studies with the bacon pig and their analysis. *Biometrika*, **30**, 16–28

Wogan, G.N., Pagliolunga, S. & Newberne, P.M. (1974) Carcinogenic effects of low dietary levels of aflatoxin B_1 in rats. *Food Cosmet. Toxicol.*, **12**, 681–685

Woolson, R.F. & Leeper, J.D. (1980) Growth curve analysis of complete and incomplete longitudinal data. *Commun. Stat.-Theor. Meth.*, **A9**, 1491–1513

World Health Organization (1984) *Guidelines for Drinking Water Quality*, Vol. I, *Recommendations*, Geneva

Yamagiwa, K. & Ichikawa, K. (1915) Experimentelle Studie über die Pathogenese der Epithelialgeschwülste. *Mitt. Med. Fak. Kais. Univ. Tokyo*, **15**, 295–344

APPENDICES

APPENDIX I

SOLUTION TO TEACHING EXAMPLE

In the artificial example given in Section 4.4, tumours develop in six out of 15 animals in group 1 (control group), eight out of 15 in group 2 (low-dose group) and eight out of 15 in group 3 (high-dose group). We shall now outline the analysis of this example, using the methods given in Chapter 5.

(a) Comparing these crude proportions according to Section 5.3 with option (3b) to choose the denominator, which leaves 14 animals at risk in group 3, results in no significant difference between any of the two groups and no significant trend with increasing dose levels (using scores 0, 1 and 2). The normal deviate of the Cochran-Armitage test for trend is 0.93 with a one-sided p-value of 0.176.

(b) Using time-to-death information as available, but considering that all tumours were found in an incidental context, leads to a prevalance analysis for nonlethal occult tumours as described in Section 5.4. Separation of the time scale into ad-hoc runs (as illustrated in Table 5.4) leads in this case, to the following intervals with increasing prevalance:

$$[80, 80]: \quad 0/1 = 0.00$$
$$[90, 120]: \quad 2/7 = 0.29$$
$$[130, 180]: \quad 14/26 = 0.54$$
$$[185, 200]: \quad 6/11 = 0.55$$

From the resulting contingency tables, one derives the following observed and expected numbers of tumours in the three groups:

control group: $O_0 = 6$; $E_0 = 8.12$

low-dose group: $O_1 = 8$; $E_1 = 7.09$

high-dose group: $O_2 = 8$; $E_2 = 6.79$

The test for trend (again using scores 0, 1, 2) according to formula (5.4) gives $X^2_{PT} = 1.61$, corrresponding to a normal deviate $Z_{PT} = 1.27$, which leads to a one-sided p-value for positive trend of 0.102.

(c) If one considers all tumours to be found in a fatal context, methods described in Section 5.5 for the analysis of rapidly lethal occult tumours and of observable tumours would be applied.

Table A.1   Analysis of artificial example using context of observation

Group	Tumours found in a fatal context		Tumours found in an incidental context		Combined	
	Observed	Expected	Observed	Expected	Observed	Expected
Control	2	4.64	4	5.17	6	9.81
Low-dose	3	3.10	5	3.84	8	6.95
High-dose	5	2.26	3	2.99	8	5.24

Observed and expected numbers are in this case:

$$\text{control group:} \quad O_0 = 6: \quad E_0 = 10.26$$

$$\text{low-dose group:} \quad O_1 = 8; \quad E_1 = 6.92$$

$$\text{high-dose group:} \quad O_2 = 8; \quad E_2 = 4.82.$$

The test for positive trend according to formula (5.4) gives here $X^2_{PT} = 4.55$, corresponding to a normal deviate $Z_{PT} = 2.13$, which results in a one-sided p-value for positive trend of 0.017.

(d)   Finally, making full use of the context of observation given in this data set, methods described in Section 5.6 would lead to the results summarized in Table A.1. Contexts coded 1 or 2 were considered as incidential tumours, and contexts 3 and 4 as fatal tumours. Animals with tumours found in the fatal context were not considered when computing the ad-hoc runs for the analysis of tumours found in an incidental context. The intervals are in this case: [80, 80], [100, 120], [130, 180], [200, 200].

Table A.1 gives the observed and expected numbers of tumours in the different contexts, both separately and combined. Calculating the trend statistic as described in Section 5.6 gives $X^2_{CT} = 3.96$, corresponding to a normal deviate $Z_{CT} = 1.99$, which results in a one-sided p-value for positive trend of 0.024.

Without presenting a full survival analysis, it can be seen directly from the data in Table 4.4 that there is increasing mortality with increasing dose. Five animals in group 1 survive to 'terminal sacrifice' at time 200, three in group 2 and one in group 3. Correction for the difference in mortality is essential for an unbiased analysis. Considering all tumours to be found in a fatal context would overestimate the true effect, considering all tumours to be found in an incidental context would not lead to a significant trend, and the result considering contexts of observation lies between these two.

APPENDIX II

COMPUTER PROGRAMS

Several computer programs are available for the computation involved in the different methods described in this book. There are general statistical packages, in the framework of which some analyses can be conducted, special programs written to deal with data from long-term animal experiments, and specific programs developed for special methods. Some of these have already been mentioned in the text. In this appendix, we mention briefly some of the computer programs that can be applied. No claim of completeness is made.

Throughout this monograph, it is apparent that the analysis of long-term animal experiments requires statistical methods that are adapted to the biological particularities of the field. Consequently, large statistical packages such as BMDP, SAS and SPSS, which address general statistical methods, may not suffice for the specific needs of the analysis of a long-term experiment. Nevertheless, the wide range of methods offered by these packages would allow the majority of statistical calculations to be performed. When they are composed appropriately, full analysis of an experiment can be carried out.

Methods for analysing censored survival data are included in all these packages and allow for the estimation and comparison of survival curves, as discussed in Section 5.3, and can also be used to analyse the occurrence of rapidly lethal occult tumours or observable tumours, as outlined in Section 5.6.

Regression methods for survival data, such as Cox's proportional hazards model, as discussed in Section 6.3, are also a common part of these packages. Methods for the analysis of crude proportions (Section 5.4) can be found in the framework of programs for the analysis of frequency or contingency tables, which also provide a tool for performing analyses of nonlethal occult tumours (Section 5.5), including the regression analyses with logistic models (see also Section 6.3). The methods for the analysis of auxiliary data outlined in Chapter 8 are mainly textbook methods. Thus, the larger statistical packages include programs for the parametric analysis of variance, both for univariate (Section 8.3) and multivariate data, and allowing, in the latter case, for the structure of repeated measurements (Section 8.4).

In any case, when using the facilities of larger statistical packages for the analysis of long-term animal experiments, careful consideration must be given to definition of the appropriate variables in order to analyse properly the correct endpoint(s).

Some special computer programs that have been published in the scientific literature are referenced at appropriate places in the text. These are basically programs for the analysis of contingency tables (Chapter 5) and parametric Weibull models for time-to-tumour data (Section 6.3). In many of the papers that give special methods, as outlined in Chapter 7, special computer programs are said to be available for the specific purpose. In Chapter 3, we referred to some computer systems that have been developed for the routine storage and management of histopathological data.

Following the publication of Peto *et al.* (1980), a computer program was developed for the analysis of occult tumours when contexts of observation are known (Section 5.7). This simple FORTRAN program, named CARTEST, is available free of charge from IARC in Lyon, France. A corresponding APL program has been published by Rosenkranz (1982), and these methods are also part of the system developed by Roe and Lee (1984).

Finally, a program named RISK81, to fit different dose-response models to crude proportions of tumour-bearing animals, as outlined in Section 6.2, is available from D. Krewski at Health and Welfare Canada in Ottawa, Canada.

SUBJECT INDEX

A

Abelson–Tukey test, 174
Acceleration
  definition, 7
  fatal tumours, 13
  incidental tumours, 14–15
  test for, with fatal tumours, 87
2-Acetylaminofluorine (2-AAF), 43, 49,
  109–10, 119
Adaptive internal selection method, 89–90
Additive model, 158
Adjustment for intercurrent mortality, 8, 13,
  15
Age-specific hazard rates, 160
Age-specific incidence rates, 127
Analysis of variance, 171–80
  nonparametric, 177
  parametric, 172
Animal carcinogenesis experiments, 6–20,
  70–106
Animal selection, 24
Armitage–Doll model, 112
Armitage test, 174
Association among tumour types, 165
Asymptotic approximations
  2×2 table, 77
  $2 \times k$ table, 82, 97
  rank tests with censored data, 71
Auxiliary data, 170

B

Background response, 111
Benzo[a]pyrene (BP), 50, 132–35
Bias
  due to intercurrent mortality, 8, 13–15
  due to lack of randomization, 30
  in analysis of fatal tumours, 13–14
  in analysis of incidental tumours, 14–15
Blind pathology, 51
Bonferroni correction, 87

Bonferroni method (*see* Multiple
  comparisons)
Breslow-modified Wilcoxon test, 101

C

Cage effects (*see* Randomization)
Carcinogenesis mechanism studies, 47–50, 54
Carcinogenic potency, estimation, 120, 121
Carcinogenic potential, 41
Carcinogenicity bioassay, 23, 27
Carcinogenicity evaluation, 19–20
Chi-squared statistics, 71, 79, 83, 97
Cigar-smoke condensate, 95–100, 129, 130,
  135
Cochran–Armitage trend test, 83, 86
Combining evidence from several
    experiments
  analysis of crude tumour rates, 75–87
  analysis of tumour incidence curves, 96–
    98
  contexts of observation, 15
  rationale, 10
Competing risks, 124–25
Computer programs, 212–13
Concomitant information, 19
Conditional maximum likelihood estimator,
  81
Conditional point estimates, 81
Confidence intervals
  nonparametric for percentile, 73
  odds ratio for crude tumour rates, 81, 82
  quantiles for low-dose extrapolation, 117
Contexts of observation, 15–17
  definition, 12
  use in analysis, 15–16, 101
Contingency tables, 70, 90
Continuity correction
  for 2×2 table, 79
  for $2 \times k$ table, 86

PUBLICATIONS OF THE INTERNATIONAL AGENCY FOR RESEARCH ON CANCER

SCIENTIFIC PUBLICATIONS SERIES

(Available from Oxford University Press)

No. 1 LIVER CANCER (1971)
176 pages; out of print

No. 2 ONCOGENESIS AND HERPES
VIRUSES (1972)
Edited by P.M. Biggs, G. de-Thé & L.N. Payne
515 pages; out of print

No. 3 N-NITROSO COMPOUNDS -
ANALYSIS AND FORMATION (1972)
Edited by P. Bogovski, R. Preussmann & E.A.
Walker
140 pages

No. 4 TRANSPLACENTAL
CARCINOGENESIS (1973).
Edited by L. Tomatis & U. Mohr,
181 pages; out of print

No. 5 PATHOLOGY OF TUMOURS IN
LABORATORY ANIMALS. VOLUME 1.
TUMOURS OF THE RAT. PART 1 (1973)
Editor-in-Chief V.S. Turusov
214 pages

No. 6 PATHOLOGY OF TUMOURS IN
LABORATORY ANIMALS. VOLUME 1.
TUMOURS OF THE RAT. PART 2 (1976)
Editor-in-Chief V.S. Turusov
319 pages

No. 7 HOST ENVIRONMENT INTER-
ACTIONS IN THE ETIOLOGY OF
CANCER IN MAN (1973)
Edited by R. Doll & I. Vodopija,
464 pages

No. 8 BIOLOGICAL EFFECTS OF
ASBESTOS (1973)
Edited by P. Bogovski, J.C. Gilson,
V. Timbrell & J.C. Wagner,
346 pages; out of print

No. 9 N-NITROSO COMPOUNDS IN
THE ENVIRONMENT (1974)
Edited by P. Bogovski & E.A. Walker
243 pages

No. 10 CHEMICAL CARCINOGENESIS
ESSAYS (1974)
Edited by R. Montesano & L. Tomatis,
230 pages

No. 11 ONCOGENESIS AND HERPES-
VIRUSES II (1975)
Edited by G. de-Thé, M.A. Epstein
& H. zur Hausen
Part 1, 511 pages
Part 2, 403 pages

No. 12 SCREENING TESTS IN
CHEMICAL CARCINOGENESIS (1976)
Edited by R. Montesano, H. Bartsch &
L. Tomatis
666 pages

No. 13 ENVIRONMENTAL POLLUTION
AND CARCINOGENIC RISKS (1976)
Edited by C. Rosenfeld & W. Davis
454 pages; out of print

No. 14 ENVIRONMENTAL N-NITROSO
COMPOUNDS — ANALYSIS AND
FORMATION (1976)
Edited by E.A. Walker, P. Bogovski &
L. Griciute
512 pages

No. 15 CANCER INCIDENCE IN FIVE
CONTINENTS. VOL. III (1976)
Edited by J. Waterhouse, C.S. Muir,
P. Correa & J. Powell
584 pages

No. 16 AIR POLLUTION AND CANCER
IN MAN (1977)
Edited by U. Mohr, D. Schmahl &
L. Tomatis
331 pages; out of print

No. 17 DIRECTORY OF ON-GOING
RESEARCH IN CANCER
EPIDEMIOLOGY 1977 (1977)
Edited by C.S. Muir & G. Wagner,
599 pages; out of print

No. 18 ENVIRONMENTAL CARCINOGENS.
SELECTED METHODS OF ANALYSIS
Editor-in-Chief H. Egan
Vol. 1. ANALYSIS OF VOLATILE
NITROSAMINES IN FOOD (1978)
Edited by R. Preussmann, M. Castegnaro,
E.A. Walker & A.E. Wassermann
212 pages; out of print

SCIENTIFIC PUBLICATIONS SERIES

No. 19 ENVIRONMENTAL ASPECTS
OF N-NITROSO COMPOUNDS (1978)
Edited by E.A. Walker, M. Castegnaro,
L. Griciute & R.E. Lyle
566 pages

No. 20 NASOPHARYNGEAL
CARCINOMA: ETIOLOGY AND
CONTROL (1978)
Edited by G. de-Thé & Y. Ito,
610 pages; out of print

No. 21 CANCER REGISTRATION
AND ITS TECHNIQUES (1978)
Edited by R. MacLennan, C.S. Muir,
R. Steinitz & A. Winkler
235 pages

No. 22 ENVIRONMENTAL CARCINOGENS.
SELECTED METHODS OF ANALYSIS
Editor-in-Chief H. Egan
Vol. 2. METHODS FOR THE MEASURE-
MENT OF VINYL CHLORIDE IN
POLY(VINYL CHLORIDE), AIR, WATER
AND FOODSTUFFS (1978)
Edited by D.C.M. Squirrell & W. Thain,
142 pages; out of print

No. 23 PATHOLOGY OF TUMOURS IN
LABORATORY ANIMALS. VOLUME II.
TUMOURS OF THE MOUSE (1979)
Editor-in-Chief V.S. Turusov
669 pages

No. 24 ONCOGENESIS AND HERPES-
VIRUSES III (1978)
Edited by G. de-Thé, W. Henle & F. Rapp
Part 1, 580 pages
Part 2, 522 pages; out of print

No. 25 CARCINOGENIC RISKS -
STRATEGIES FOR INTERVENTION
(1979)
Edited by W. Davis & C. Rosenfeld,
283 pages; out of print

No. 26 DIRECTORY OF ON-GOING
RESEARCH IN CANCER EPI-
DEMIOLOGY 1978 (1978)
Edited by C.S. Muir & G. Wagner,
550 pages; out of print

No. 27 MOLECULAR AND CELLULAR
ASPECTS OF CARCINOGEN
SCREENING TESTS (1980)
Edited by R. Montesano, H. Bartsch & L. Tomatis
371 pages

No. 28 DIRECTORY OF ON-GOING
RESEARCH IN CANCER EPI-
DEMIOLOGY 1979 (1979)
Edited by C.S. Muir & G. Wagner,
672 pages; out of print

No. 29 ENVIRONMENTAL CARCINOGENS.
SELECTED METHODS OF ANALYSIS
Editor-in-Chief H. Egan
Vol. 3. ANALYSIS OF POLYCYCLIC
AROMATIC HYDROCARBONS IN
ENVIRONMENTAL SAMPLES (1979)
Edited by M. Castegnaro, P. Bogovski,
H. Kunte & E.A. Walker
240 pages; out of print

No. 30 BIOLOGICAL EFFECTS OF
MINERAL FIBRES (1980)
Editor-in-Chief J.C. Wagner
Volume 1, 494 pages
Volume 2, 513 pages

No. 31 N-NITROSO COMPOUNDS:
ANALYSIS, FORMATION AND
OCCURRENCE (1980)
Edited by E.A. Walker, M. Castegnaro,
L. Griciute & M. Börzsönyi
841 pages; out of print

No. 32 STATISTICAL METHODS IN
CANCER RESEARCH
Vol. 1. THE ANALYSIS OF CASE-
CONTROL STUDIES (1980)
By N.E. Breslow & N.E. Day
338 pages

No. 33 HANDLING CHEMICAL
CARCINOGENS IN THE LABORATORY -
PROBLEMS OF SAFETY (1979)
Edited by R. Montesano, H. Bartsch,
E. Boyland, G. Della Porta, L. Fishbein,
R.A. Griesemer, A.B. Swan & L. Tomatis,
32 pages

No. 34 PATHOLOGY OF TUMOURS
IN LABORATORY ANIMALS. VOLUME
III. TUMOURS OF THE HAMSTER
(1982)
Editor-in-Chief V.S. Turusov,
461 pages

No. 35 DIRECTORY OF ON-GOING
RESEARCH IN CANCER EPIDEMIOLOGY
1980 (1980)
Edited by C.S. Muir & G. Wagner,
660 pages; out of print

SCIENTIFIC PUBLICATIONS SERIES

No. 36 CANCER MORTALITY BY
OCCUPATION AND SOCIAL CLASS
1851-1971 (1982)
By W.P.D. Logan
253 pages

No. 37 LABORATORY DECONTAMI-
NATION AND DESTRUCTION OF
AFLATOXINS B_1, B_2, G_1, G_2 IN
LABORATORY WASTES (1980)
Edited by M. Castegnaro, D.C. Hunt,
E.B. Sansone, P.L. Schuller,
M.G. Siriwardana, G.M. Telling,
H.P. Van Egmond & E.A. Walker,
59 pages

No. 38 DIRECTORY OF ON-GOING
RESEARCH IN CANCER EPI-
DEMIOLOGY 1981 (1981)
Edited by C.S. Muir & G. Wagner,
696 pages; out of print

No. 39 HOST FACTORS IN HUMAN
CARCINOGENESIS (1982)
Edited by H. Bartsch & B. Armstrong
583 pages

No. 40 ENVIRONMENTAL CARCINOGENS
SELECTED METHODS OF ANALYSIS
Editor-in-Chief H. Egan
Vol. 4. SOME AROMATIC AMINES AND
AZO DYES IN THE GENERAL AND
INDUSTRIAL ENVIRONMENT (1981)
Edited by L. Fishbein, M. Castegnaro,
I.K. O'Neill & H. Bartsch
347 pages

No. 41 N-NITROSO COMPOUNDS:
OCCURRENCE AND BIOLOGICAL
EFFECTS (1982)
Edited by H. Bartsch, I.K. O'Neill,
M. Castegnaro & M. Okada,
755 pages

No. 42 CANCER INCIDENCE IN FIVE
CONTINENTS. VOLUME IV (1982)
Edited by J. Waterhouse, C. Muir,
K. Shanmugaratnam & J. Powell,
811 pages

No. 43 LABORATORY DECONTAMI-
NATION AND DESTRUCTION OF
CARCINOGENS IN LABORATORY
WASTES: SOME N-NITROSAMINES
(1982) Edited by M. Castegnaro,
G. Eisenbrand, G. Ellen, L. Keefer,
D. Klein, E.B. Sansone, D. Spincer,
G. Telling & K. Webb
73 pages

No. 44 ENVIRONMENTAL CARCINOGENS.
SELECTED METHODS OF ANALYSIS
Editor-in-Chief H. Egan
Vol. 5. SOME MYCOTOXINS (1983)
Edited by L. Stoloff, M. Castegnaro,
P. Scott, I.K. O'Neill & H. Bartsch,
455 pages

No. 45 ENVIRONMENTAL CARCINOGENS.
SELECTED METHODS OF ANALYSIS
Editor-in-Chief H. Egan
Vol. 6: N-NITROSO COMPOUNDS
(1983)
Edited by R. Preussmann, I.K. O'Neill,
G. Eisenbrand, B. Spiegelhalder &
H. Bartsch
508 pages

No. 46 DIRECTORY OF ON-GOING
RESEARCH IN CANCER EPI-
DEMIOLOGY 1982 (1982)
Edited by C.S. Muir & G. Wagner,
722 pages; out of print

No. 47 CANCER INCIDENCE IN
SINGAPORE (1982)
Edited by K. Shanmugaratnam, H.P. Lee
& N.E. Day
174 pages; out of print

No. 48 CANCER INCIDENCE IN
THE USSR Second Revised
Edition (1983)
Edited by N.P. Napalkov,
G.F. Tserkovny, V.M. Merabishvili,
D.M. Parkin, M. Smans & C.S. Muir,
75 pages

No. 49 LABORATORY DECONTAMI-
NATION AND DESTRUCTION OF
CARCINOGENS IN LABORATORY
WASTES: SOME POLYCYCLIC
AROMATIC HYDROCARBONS (1983)
Edited by M. Castegnaro, G. Grimmer,
O. Hutzinger, W. Karcher, H. Kunte,
M. Lafontaine, E.B. Sansone, G. Telling
& S.P. Tucker
81 pages

No. 50 DIRECTORY OF ON-GOING
RESEARCH IN CANCER EPI-
DEMIOLOGY 1983 (1983)
Edited by C.S. Muir & G. Wagner,
740 pages; out of print

SCIENTIFIC PUBLICATIONS SERIES

SCIENTIFIC PUBLICATIONS SERIES

NON-SERIAL PUBLICATIONS

(Available from IARC)

ALCOOL ET CANCER (1978)
By A.J. Tuyns (in French only)
42 pages

CANCER MORBIDITY AND CAUSES OF
DEATH AMONG DANISH BREWERY
WORKERS (1980)
By O.M. Jensen
145 pages

DIRECTORY OF COMPUTER SYSTEMS
USED IN CANCER REGISTRIES (1986)
By H.R. Menck & D.M. Parkin
236 pages

IARC MONOGRAPHS ON THE EVALUATION OF THE CARCINOGENIC RISK OF CHEMICALS TO HUMANS

(English editions only)

(Available from WHO Sales Agents)

Volume 1
Some inorganic substances, chlorinated hydrocarbons, aromatic amines, N-nitroso compounds, and natural products (1972)
184 pp.; out of print

Volume 2
Some inorganic and organometallic compounds (1973)
181 pp.; out of print

Volume 3
Certain polycyclic aromatic hydrocarbons and heterocyclic compounds (1973)
271 pp.; out of print

Volume 4
Some aromatic amines, hydrazine and related substances, N-nitroso compounds and miscellaneous alkylating agents (1974)
286 pp.

Volume 5
Some organochlorine pesticides (1974)
241 pp.; out of print

Volume 6
Sex hormones (1974)
243 pp.

Volume 7
Some anti-thyroid and related substances, nitrofurans and industrial chemicals (1974)
326 pp.; out of print

Volume 8
Some aromatic azo compounds (1975)
357 pp.

Volume 9
Some aziridines, N-, S- and O-mustards and selenium (1975)
268 pp.

Volume 10
Some naturally occurring substances (1976)
353 pp.; out of print

Volume 11
Cadmium, nickel, some epoxides, miscellaneous industrial chemicals and general considerations on volatile anaesthetics (1976)
306 pp.

Volume 12
Some carbamates, thiocarbamates and carbazides (1976)
282 pp.

Volume 13
Some miscellaneous pharmaceutical substances (1977)
255 pp.

Volume 14
Asbestos (1977)
106 pp.

Volume 15
Some fumigants, the herbicides 2,4-D and 2,4,5-T, chlorinated dibenzodioxins and miscellaneous industrial chemicals (1977)
354 pp.

Volume 16
Some aromatic amines and related nitro compounds - hair dyes, colouring agents and miscellaneous industrial chemicals (1978)
400 pp.

Volume 17
Some N-nitroso compounds (1978)
365 pp.

Volume 18
Polychlorinated biphenyls and polybrominated biphenyls (1978)
140 pp.

Volume 19
Some monomers, plastics and synthetic elastomers, and acrolein (1979)
513 pp.

Volume 20
Some halogenated hydrocarbons (1979)
609 pp.

Volume 21
Sex hormones (II) (1979)
583 pp.

Volume 22
Some non-nutritive sweetening agents (1980)
208 pp.

Volume 23
Some metals and metallic compounds (1980)
438 pp.

Volume 24
Some pharmaceutical drugs (1980)
337 pp.

Volume 25
Wood, leather and some associated industries (1981)
412 pp.

Volume 26
Some antineoplastic and immuno-suppressive agents
(1981)
411 pp.

Volume 27
*Some aromatic amines, anthraquinones and nitroso
compounds, and inorganic
fluorides used in drinking-water and
dental preparations* (1982)
341 pp.

Volume 28
The rubber industry (1982)
486 pp.

Volume 29
Some industrial chemicals and dyestuffs (1982)
416 pp.

Volume 30
Miscellaneous pesticides (1983)
424 pp.

Volume 31
*Some food additives, feed additives and
naturally occurring substances* (1983)
314 pp.

Volume 32
*Polynuclear aromatic compounds,
Part 1, Environmental and experimental data* (1984)
477 pp.

Volume 33
*Polynuclear aromatic compounds, Part 2, Carbon
blacks, mineral oils and some nitroarene compounds*
(1984)
245 pp.

Volume 34
*Polynuclear aromatic compounds,
Part 3, Industrial exposures
in aluminium production, coal gasification, coke
production, and iron and steel founding* (1984)
219 pp.

Volume 35
*Polynuclear aromatic compounds,
Part 4, Bitumens, coal-tar and
derived products, shale-oils and soots*
(1985)
271 pp.

Volume 36
*Allyl Compounds, aldehydes,
epoxides and peroxides* (1985)
369 pp.

Volume 37
*Tobacco habits other than smoking; betel-quid and
areca-nut chewing;
and some related nitrosamines* (1985)
291 pp.

Volume 38
Tobacco smoking (1986)
421 pp.

Volume 39
Some chemicals used in plastics and elastomers
(1986)
403 pp.

Volume 40
*Some naturally occurring and synthetic food
components, furocoumarins and ultra-violet
radiation* (1986)
444 pp.

Volume 41
*Some halogenated hydrocarbons and pesticide
exposures* (1986)
434 pp.

Supplement No. 1
*Chemicals and industrial processes associated with
cancer in humans (IARC Monographs,
Volumes 1 to 20)* (1979)
71 pp.; out of print

Supplement No. 2
*Long-term and short-term screening assays
for carcinogens: a critical appraisal* (1980)
426 pp.; US$25.00; Sw.fr. 40.-

Supplement No. 3
*Cross index of synonyms and trade
names in Volumes 1 to 26* (1982)
199 pp.

Supplement No. 4
*Chemicals, industrial processes and
industries associated with cancer in humans
(IARC Monographs, Volumes 1 to 29)* (1982)
292 pp.

Supplement No. 5
*Cross index of synonyms and trade
names in Volumes 1 to 36* (1985)
259 pp.

INFORMATION BULLETINS ON THE
SURVEY OF CHEMICALS BEING
TESTED FOR CARCINOGENICITY

(Available from IARC)

No. 8 (1979)
Edited by M.-J. Ghess, H. Bartsch
& L. Tomatis
604 pp.; US$20.00; Sw.fr. 40.-

No. 9 (1981)
Edited by M.-J. Ghess, J.D. Wilbourn,
H. Bartsch & L. Tomatis
294 pp.; US$20.00; Sw.fr. 41.-

No. 10 (1982)
Edited by M.-J. Ghess, J.D. Wilbourn
H. Bartsch
326 pp.; US$20.00; Sw.fr. 42.-

No. 11 (1984)
Edited by M.-J. Ghess, J.D. Wilbourn,
H. Vainio & H. Bartsch
336 pp.; US$20.00; Sw.fr. 48.-

No. 12 (1986)
Edited by M.-J. Ghess, J.D. Wilbourn,
A. Tossavainen & H. Vainio
389 pp.